THE DEVELOPMENT OF THE BRAZILIAN STEEL INDUSTRY

The
Development of the
Brazilian
Steel Industry

WERNER BAER

Vanderbilt University Press
1969

*To Miguel Colasuonno, Francisco Camargo,
Isaac Kerstenetzky, Mario H. Simonsen,
leaders in the modernization of
Brazil's economics profession.*

PREFACE

This study was made possible by a grant from the Social Science Research Council in the academic year 1965–66, which enabled me to do the necessary field work in Brazil. My deep gratitude also goes to the Instituto Brasileiro de Economia of the Fundação Getúlio Vargas, which provided me with a research home and gave me the cover of its great prestige in making contacts in the steel industry. Special thanks go to Julian M. Chacel, its executive director, Isaac Kerstenetzky, its research director, José Almeida, the director of its industrial research sector, and Mario H. Simonsen, the director of its graduate school, for providing encouragement and stimulation. Most of the writing of this volume was done in the Instituto de Pesquisas Econômicas of the University of São Paulo, which provided me with office space, all needed equipment, and a stimulating atmosphere.

I am greatly indebted to all the major and many of the smaller steel companies with which I had direct contact. Both private and government firms were extremely co-operative in providing information and extremely generous in the hospitality they showed during my visits. I must also mention officials of the Brazilian National Development Bank (BNDE), the Development Bank of Minas Gerais, and the private consulting firm Tecnometal, all of whom were very generous in their co-operation.

Although space does not permit the listing of the large number of people who have helped me, I would like to mention a few who went out of their way in co-operating. Wando Borges of the University of Minas

Gerais, currently with the Ministry of Transport, went beyond the call of duty of a colleague and friend in accompanying me on visits to many steel plants in Minas Gerais, in making contacts, in collecting data for me, and in helping me interpret. Marcos Contrucci of Tecnometal and the Catholic University was one of my indefatigable teachers of steel technology and always ready to provide me with crucial references. Colonel Cyro Alves Borges of the Companhia Siderúrgica Nacional, who runs one of the most efficient statistical services in Latin America, was always ready to put his vast resources at my disposal and always cheerfully ready to let me pick his brain. Marcio Augusto de Menezes, formerly with the Development Bank of Minas Gerais and currently with Tecnometal, was also most helpful in letting me benefit from his vast experience in steel technology and its application in Brazil. I would also like to acknowledge the valuable help received from F. S. H. Pegurier, Aluísio Marins, Joel Bergman, Hans Mueller, and Peter Kilby.

Carlos Pelaez, working on a study of Brazil's pre–World War II industrialization, was most generous in showing me his findings. His sharp analytical mind and his intellectual independence made him a valuable critic as my work proceeded.

Last but not least, let me mention two friends who have been most valuable in giving me both technical and moral support. First, José Mariano Falcão, one of Brazil's outstanding steel authorities, who not only taught me a lot about steel technology, but also about what is involved in implanting steel mills in developing countries. He also contributed generously of his time in reading and criticizing my work. Second, my thanks go to William O. Thweatt, who is not a steel expert but who is one of those few economists who combine a first-rate analytical mind with common sense. He was most generous in letting me use him constantly to test out my ideas.

Special thanks also go to my patient typist, Zulmira Alessandrini, who attacked some of the more monstrous tables of this volume without a murmur.

Needless to say, none of the above-mentioned individuals is responsible for any weaknesses in the final product.

WERNER BAER

CONTENTS

TABLES

THE DEVELOPMENT OF THE BRAZILIAN STEEL INDUSTRY

1

INTRODUCTION

THE ECONOMIC development of Brazil since the Second World War has
been based on industrial growth. While the real gross domestic product
grew at an annual rate of nearly six percent in the period from the late
forties to the early sixties, industrial production grew at an annual rate of
nearly ten percent. The government's import substitution policies were
the main impulse behind the industrialization spurt. From the very beginning, one of the chief characteristics of these policies has been the
stimulation of vertical integration within the developing industrial complex; that is, the government tried to stimulate a maximum amount of
forward and backward linkages within the economy,[1] thus adding a maximum value of industrial production within the country. Consequently,
the establishment and expansion of both light and heavy industries simultaneously can be observed throughout this period.

The Brazilian steel industry, which had already made considerable
progress in the interwar period, was thus greatly affected by the government's approach to the country's industrial growth. Steel has been one of
the more dynamic sectors in Brazil's industrial complex since the Second
World War. Steel output grew at an annual rate more than 50 percent
higher than that of industrial production.

Two fundamental reasons exist for writing a specialized study of the
Brazilian steel industry.

First, economists seem to have come to a stage where few insights into
the process of economic development can be gained by aggregative analysis. The latter often raises more questions than it answers. For example,

1. For a detailed description and analysis of this policy see Werner Baer, *Industrialization and Economic Development in Brazil,* Chapters Three and Six.

an analysis of the Brazilian economy since the Second World War reveals a low capital-output ratio and a low rate of labor absorption by the industrial sector. Since no sectoral investment data are available, little is known about capital productivity in various sectors and little can be said about the reasons for the low over-all capital-output ratio. Also, little effort has been made to study the economics of technology in different industries; thus, little that is definitive can be said about the reasons for the low rate of labor absorption in industry. Only a series of industry studies can help us to find an answer to these problems.

Another set of questions which can only be answered by individual industry studies relates to the process of implanting industries into a formerly rural and backward economy. How fast do factors of production adapt to new technology? How fast can costs be brought down to competitive levels? How long does the new industry have to be protected? To what extent is the dearth of skilled manpower and managerial personnel a bottleneck for new industries?

I have chosen to study the Brazilian steel industry, not only to get some insights into such problems as those stated above, but also because it represents an effort to establish a heavy industry in a developing economy. Many who are skeptical of industrialization as the best path to economic development are especially critical of countries not satisfied with the development of light industries, and which thus encourage the establishment of steel enterprises and various types of capital goods industries. For example, H. G. Johnson states that

nationalism derives its values and objectives from rivalry with, and imitation of, other and better-established nations. One result is considerable diversion of economic resources from productive investment in economic development to consumption of the trappings and symbols of nationhood—a large and well-equipped army, an elaborate diplomatic bureaucracy, impressive public buildings, and other constructional monuments to national pride. Another result is the shaping of investment plans by a desire for the industrial structure considered essential to a large and important nation—a steel industry, an automotive industry—rather than by considerations of maximum profitability of investment in the circumstances of time and place.[2]

Other economists who accept some degree of industrialization as a path to increased economic growth feel that such industrialization should be restricted to the development of light industries because only their products might have a place in the international market. A good example is G. M. Meier's view:

2. Harry G. Johnson, *The World Economy at the Crossroads*, pp. 77–79.

Beyond the benefits from improvements in the position of their traditional primary exports, the poor countries may be able to take advantage of new export opportunities for manufactured goods. The exportation of cheap labor-intensive manufactured commodities may provide an increasingly important opportunity for transmitting development to some poor countries. . . . It may be only through production for export that a poor country can overcome the small size of its home market and become efficient in industries that need a wide market in order to realize economies of scale. It is unlikely, however, that capital goods or the products of heavy industry can be successfully exported because they require considerable capital and large-scale methods.[3]

In other words, steel mills and other heavy industries are thought to be outside the range of development activities of poor countries, since they are too complex for the existing labor and managerial talents available, need too much capital and require too large a market in relation to the low saving levels and small domestic demand.

I have long suspected that these generalizations do not necessarily apply to all developing countries. The very fact that different developing countries have different types of factor endowments, and are of different geographic and population sizes, leads one to distrust all-inclusive theories about legitimate spheres of economic activity of these countries.

In this book I shall review the growth of the Brazilian steel industry, its effect on the growth of the economy, and its performance. There are a number of reasons for choosing this industry rather than other heavy industries in Brazil. First, it is a large and important industry. With its output of almost 3.8 million ingot tons of steel in 1966, it was the largest steel industry in Latin America and with expansion plans which will probably lead to output levels of 10 million tons in the mid 1970s, it was most likely to remain the most important steel industry in Latin America. Second, the industry has a longer history than most other heavy industries in Brazil. Some firms have existed for more than forty years, and Volta Redonda, Latin America's first coke-based integrated steel mill, has been functioning for more than 20 years. Given this length of time, a more definitive analysis may be made of the adaptation of factors of production to the technology, of technology to factor availability, and of changes in the relative efficiency of the industry. That is, a long enough time has passed for the economist to be able to make some judgments about the extent to which the infant industry had grown up. Third, both the private (domestic and foreign) and the government sectors were and are involved directly in building up the industry; this provides a unique opportunity for studying

3. Gerald M. Meier, *International Trade and Development,* pp. 189-190.

the way both the private and the public sector are able to marshall and administer resources in organizing new productive units.

In the next chapter, we begin with a description of steel-making technology and the economics of this technology; considerable attention will be given to investment costs of various technologies, based especially on the Brazilian experience. The following chapter briefly surveys the natural resource endowment of Brazil for steel-making.

Chapter Four tells the history of the growth of the Brazilian steel industry. In Chapter Five we analyze the impact of the industry's growth on the economy, comparing the growth of steel output and the composition of this output with the growth of the economy, examining the pattern of forward linkages of the industry and its backward linkage effect. The latter consists principally of an analysis of the industry's influence on employment and its reaction to and effect on the country's manpower supply. Chapters Six and Seven contain an analysis of the industry's performance. First, an analysis of the industry's productivity, cost of production, and price formation, followed by a study of the locational pattern of the industry and the opportunity cost the industry represented for the Brazilian economy. Chapter Eight evaluates the future growth prospects of the industry, and the concluding chapter attempts to draw some general conclusions and systematize some of the new insights on the industrialization process gained by this industry study.

A word of caution is in order at this stage about the nature of our statistics. Data on investment costs and on cost of production were not readily available. They were gathered directly or indirectly from individual firms and from such government organs as the National Development Bank (BNDE). The amount and quality of information received from these various entities differed according to each firm's policies concerning the amount of information to be divulged and also according to information availability (since a number of firms did not keep systematic records of inputs or costs over the years). Thus cost information, or data from which cost estimates could be constructed, was limited to one period of time, the mid-sixties. It was therefore impossible to attempt a time-series analysis of cost trends. Some ideas of changes over time was obtained through the examination of individual productivity indicators. A systematic cross-section analysis was also impossible because of the variability of information obtained for various firms. Where possible, comparisons were made between various types of firms and with similar firms in other countries. Given the nature of the information available, it was not possible, for example, to estimate a statistical production function for the industry. Because of direct information I received on investment costs from some firms and the direct cost estimates I made for different types of firms, I was, however, able to obtain substantially richer and deeper information on capital and labor than would have come through a statistical production function.

I do not mean the above to be an apology for what is to follow. In theory, there are many things a reader of an industry study would like to know. It is thus only fair to warn him that many of the questions left unanswered or only partially answered reflect on the data available. Of course, the reader experienced in industry studies will realize more than others the limitation of information confronting the researcher and will thus also appreciate the uniqueness of some of the data I was able to obtain.

2

STEEL-MAKING:
ITS TECHNICAL NATURE AND ECONOMICS

SINCE THE purpose of an industry study is to discover to what extent it is meaningful to talk about alternative techniques of production with different factor mixes and lumpiness of investment, some knowledge of technology is indispensable. After a description of steel-making technology, I shall discuss briefly what types of technological choices are available to the planner. Finally, I shall examine the cost of building a steel mill, using some of the data I was able to obtain from a number of Brazilian steel firms.

STEEL TECHNOLOGY

In this section I shall describe only the functioning of an integrated steel plant.[1] Although the Brazilian steel industry includes many semi-integrated plants, their departments are equivalent to those of an integrated steel plant.

The production of final steel products involves four basic stages of production: the mining and treatment of raw materials, the reduction of iron ore to pig iron, the transformation (or refining) of pig iron into steel, and the rolling of steel ingots into final steel products. Although these are the basic stages of production through which steel products pass, a number of important ancillary activities are essential to the functioning of these basic production stages. This will become obvious presently.

---

1. The most important reference work on steel-making is *The Making Shaping and Treating of Steel,* edited by Harold E. McGannon. For the nonprofessional, a less weighty and simpler volume is *The Making of Steel,* American Iron and Steel Institute.

Raw-Material Preparation

The basic raw material (nonlabor and noncapital) inputs of an integrated steel plant are iron ore, coal, scrap, ferro alloys, fluxes (mainly limestone). The other principal inputs are fuel oil, natural gas, oxygen, refractories, water and electric power.

Iron may be mined by either underground or open-pit methods. In Brazil the latter is used. When ores arrive at the steel mill, they are either directly charged to the blast furnace or they may at first go through a process of "beneficiation," a process improving the iron-bearing minerals through blending and sizing before use in the furnace. In many Brazilian firms the process of "sintering" is used before charging a blast furnace. A sintering machine uses fine iron ores and fine iron-bearing particles mixed with coke fines and fluxes and spread on the moving bed of the machine. The layer of particles fuses into a cake, which is broken into pieces. The use of sinter substantially lowers the requirement of coke consumption in the blast furnace.

Another preparation of iron ore, "pelletizing," has not been used up to the mid-sixties in Brazil. At present, a two-million-ton-per-year plant is being erected, with an output destined exclusively for export.

It is indispensable to have a reducing agent in order to extract the metal from the ore. Coal is converted into coke before its use in the blast furnace. Before reaching the coke oven, coal often has to be washed to improve its properties. This usually occurs near the coal mines. Washing of Brazil's coal is especially important, since its ash content is very high.

The coke often represents an important capital installation in an integrated steel firm. In a battery of a coke oven, gas burning in flues in the wall heats the coal to 2,000 degrees Fahrenheit. This drives off gas and tar, which are further processed into important byproducts.[2] In Brazil, these byproducts are sold directly by the steel mills, and the credit of the sales allows for a low production cost of coke. Coke being porous, unlike coal, burns inside as well as outside and doesn't fuse into a sticky mass. Coke holds up under a charge of iron and limestone fed into the blast furnace.

Charcoal may be used in the blast furnace instead of coke. Charcoal is seldom used in most great steel-making countries today, but Brazil still makes a substantial proportion of its iron in charcoal-burning blast furnaces (about one third of production in 1965). The Companhia Siderúr-

2. Tar, ammonium sulfate, benzone, toluene, naphthalene and phenol are the main products which can be turned into a great number of other chemical compounds.

gica Belgo Mineira, Brazil's fourth largest steel producing firm, uses only charcoal in its blast furnaces and is considered the largest integrated steel firm in the world using charcoal as a reducing agent. The use of charcoal makes expensive investments in coke ovens unnecessary but requires large forest reserves or tree plantations (especially eucalyptus trees) and cheap labor in the countryside for the cutting down of trees and burning the wood into charcoal. These requirements have existed in Brazil and made the existence of a large number of charcoal-using blast furnaces possible.

The future growth of charcoal-based steel production in Brazil is limited, however. With the natural wood supply close to the firms dwindling, the possibility of increasing labor costs and higher costs of land (which would have to be used for eucalyptus plantations), the cost advantages of charcoal-based production will vanish in the not too distant future. For firms with no plantations the cost of charcoal is already increasing, since the national forests are farther and farther away from the mills.

Another cost disadvantage is that blast furnaces using charcoal are limited in size. Charcoal cannot withstand the large physical load pressures of a modern large-scale blast furnace. Thus substantial scale economies have to be foregone.

The Reduction of Iron Ore to Pig Iron

The first step in the conversion of iron ore into steel occurs in the blast furnace where iron ore is converted into pig iron. Ore, coke, and fluxes are charged into the top of the blast furnace and a large amount of heated air is blown (injected in the sides at the bottom) up through the descending materials. Coke serves as both fuel and as a reducing agent, and the fluxes (the most important of which is limestone) react with impurities in the ore and thus separate them from the iron ore. The melting iron separates from the other materials and settles at the bottom, while the slag (consisting of the other materials in molten form) stays on the top of the molten iron. Many enterprises also inject natural gas, oil, and oxygen to increase temperature and speed up the melting process. The molten pig iron is tapped at the bottom of the furnace, as is the molten slag. Little in the blast furnace is wasted. The gas produced is expelled at the top of the furnace then flows through a number of heat exchangers (cowpers) where it heats the incoming blast. Only approximately one third of the blast furnace gas is necessary, however, to heat the incoming blast; the rest of the gas may be used in other parts of the plant.

The hot metal may either be cast into solid pig iron and sold to other companies or fed directly into a hot metal mixer (a huge drum lined with refractory), which brings it to the next major section of the mill, the steel melt shop.

By the early sixties, it was considered standard that to produce one ton of pig iron the requirements of a blast furnace were about 1.6 net tons of iron ore (with a major portion in the agglomerated form), 0.65 tons of coke, 0.2 tons of limestone, 0.05 tons of miscellaneous iron and steel scrap and about 4 to 4.5 tons of air.[3]

The use of agglomerated ore (sinter or pellets) and the injection of oil or oxygen can lower the coke rate substantially (consumption of coke per ton of pig iron produced) and thus greatly increase the productivity of the blast furnace.

The input of electric power into the blast furnace section depends on the size and the degree of automation of the blast furnace. The more modern the blast furnace, the more will probably be the use of automatic equipment to charge the furnace, thus increasing the use of electric power. Refractory bricks are the chief material used to construct all furnaces; they are also used in lining the vessels and ladles of a steel mill. Bricks are consumed periodically in the blast furnace refractory for the relining of the entire furnace, but this happens only every 5 to 7 years.

The Making of Steel

The principle of steel-making is the oxidation of the impurities which are contained in pig iron.

The principal steel-making furnaces used in most countries are the open hearth (Siemens Martin) process, the electric steel furnace, the BOF (basic oxygen furnace), as labeled in the U. S., known in other parts of the world as the LD Process and the Bessemer process.

The Open Hearth (SM) process is called "open" because the charge is exposed to flames sweeping over its surface. Burning fuel oil, tar, or gas cause the flames to surge over the charge from the sides of the furnace walls. The charge consists of molten pig iron, scrap, ferroalloys, iron ore, and limestone and is contained in a large, shallow furnace.

Any proportion of scrap and molten pig iron may be used in the metallic charge. For example, there is the 50–50 charge (50 percent molten pig iron and 50 percent scrap), which includes also small amounts of ferro-

---

3. In addition to pig iron, the furnace yields about 700 pounds of slag and about 6 tons of gas per ton of pig iron. Air constitutes about one half of the materials entering the furnace and gases more than ¾ leaving the furnace.

alloys for the production of different types of steel. Or, there is the "high molten pig iron practice," using 55 to 80 percent of molten pig iron. The proportions of scrap to pig iron depend on the availability of scrap. In general, however, the more scrap used, the better, since scrap is cheaper than pig iron. But scrap is often a relatively scarce commodity. Fluxes are used, largely lime and limestone, for the formation of slag, which will contain all the impurities to be removed from the liquid metal.

On the average, a "heat" (time from one charge to another) is 8 to 11 hours. This time may be reduced by injecting oxygen through lances through the roof of the furnace.

The steel is poured into a ladle (lined with refractory), and the ladle is lifted by crane to the pouring platform, where the steel is poured into ingot molds.

Most of the new steel plants in Brazil have installed LD Converters. The initials stand for two Austrian steelworks at Linz and Donawitz, where the process was first operated on a commercial scale. The process consists of charging scrap, molten pig iron, and fluxes into the top of a tilted vessel (a lined converter). The vessel is then straightened and pure oxygen is blown in from the top by means of water-cooled lances. Oxygen, which combines with carbon and other unwanted elements, raises the temperature, causing reactions that burn out impurities from the molten pig iron, converting it into steel.

One of the advantages of the LD process is the short tap-to-tap heat cycle, which takes only approximately forty-five minutes. The process uses a larger amount of hot pig iron than does the SM method; scrap content is limited to about 10 to 30 percent of the charge, depending on the size of the vessel. A plant using this process has at least two vessels, one of which is used currently while the other is being relined with refractory. After the steel is refined, the converter is tapped, necessary chemical adjustments are made, the converter is tilted, and the steel is poured into ladles and then to ingot molds.

The Bessemer process, which has been surpassed by the LD method, has until recently been used by only one Brazilian steel plant.[4] It consists of an acid refractory lined converted, which can be tilted to receive the charge. The converter has a double bottom which forms an air chamber. Air is blown through the bottom. The air pressure applied is enough to

4. The firm ACESITA in Minas Gerais has used this process. Plans are underway, however, to convert the Bessemer furnace into an LD system.

support the molten charge and to force streams of air through the metal bath.

Also commonly found in Brazil is the basic electric-arc furnace. Electric furnaces are usually charged almost exclusively with scrap. The furnace consists of a steel shell lined with refractory brick. The furnace is set on huge rockers in order to tilt it to pour out the molten metal and slag. A modern electric furnace has a removable roof so that charging can take place from above. This roof is domelike (composed of refractory brick) and through it usually three large cylindrical electrodes of carbon or graphite reach into the furnace. These electrodes carry the current to the steel charge. Electricity is used for heat, which comes from the proximity of the electric arc and the electrical resistance of the steel bath itself.

The electric furnace can produce all types of carbon and alloy steel, since the heat of the furnace is very intense, can be rigidly controlled, and is independent of any needs of oxygen for combustion. Oxygen may also be injected in order to reduce the time of the heat. Electric furnaces are used especially by firms producing special steel products.

In Brazil, electric furnaces may be found in many semi-integrated steel firms. Firms using them exclusively do not need blast furnaces and all their accessories. One also finds integrated firms using them in order to produce certain lines of special steel products or to produce their own spare parts.

The Shaping of Final Steel Products

Molten steel is usually poured into ingot molds of various shapes and sizes. These molds are made of cast iron.[5] The steel cools and hardens in the mold. It cools from the outside in, thus first hardening on the outside while remaining liquid on the inside for a longer period. After the necessary solidification the ingots are removed by cranes from their molds and placed into soaking pits to give the ingots a uniform temperature. These soaking pits are heated, much like the open hearth furnaces, by oil or gas.

From the soaking pits the ingots are lifted by crane and go to the rolling mill where they are rolled into shapes called blooms or slabs. A bloom is generally square or rectangular in cross section and is used for making

---

5. These molds are often produced by the firms themselves in their foundries; other firms buy their molds from smaller independent foundries.

nonflat products. A slab is generally wider and thinner than a bloom and is used for making flat products.

Nonflat Products:

In the blooming or slabbing mill, the ingot is passed between two huge steel rollers, passing back and forth as the rollers are brought closer together. The product is sheared into shorter lengths while still red hot. A blooming mill produces blooms which are later delivered to billet mills.

Billets, which are produced from blooms, are generally square in cross-section and usually range in size from 2 by 2 to 8 by 8 inches, depending on the size of the bloom from which they were produced. Billets are steel in an intermediary form. There are two basic types of billet mills: in the three-high mill the bloom goes back and forth through passes of different size; in the continuous-billet mill it is rolled simultaneously in several stands placed in tandem, thus speeding the rolling process. In more primitive billet mills to be found in older Brazilian firms, one still finds three-high mills where the shape is passed back and forth with tongues by workers.[6]

Billets are used to produce bars, wire products, tubes, and items of various shapes. For producing bars, the billets are placed into reheating furnaces. To produce beams or rails, the billet is then passed through rolls of various shapes until the desired shape is attained. The production of wire is achieved by a process of a cold drawing machine. Round bars, having been given a coating of lime, are passed through a machine which reduces the ends of the bar and pull the bar through a die of a drawing machine. The number of times required to draw a bar depends on the type of steel and wire desired.

Another major steel product is pipes. The two basic types that may be produced are seamless and welded pipes. The former may be produced by hot piercing: round billets are rolled through a piercing mill containing a pair of skewed double tapered rolls, forming a continuous tube from solid steel by working it over a piercing point. Welding consists of taking a plate of steel, rolling and pressing it into a hollow cylinder, then welding the seam where the edges of the plate meet. After this, the ends of the cylinder are expanded and plugged, and water is pumped through under

6. Continuous casting, used in one plant in Brazil, consists of pouring the molten metal directly into the mold. The ingot molds, soaking pits, and primary mills are bypassed. The method makes possible great savings in capital costs.

pressure, while retainer rings encircle the pipe to keep it from expanding. Welding is used for larger pipes.

Flat Products.

The slabbing mill produces a slab, which is delivered to the plate or hot-strip mill. These primary rolling mills, for large-sized capacities, are usually of the two-high reversable type. The rolling action is provided by two rollers which reverse their rotation after each pass.

Slabs are turned into plate sheet and strip steel, which are used for such products as automobiles, kitchen cabinets, and office furniture. After the slabs have been scarfed over the top and bottom surfaces, they are re-heated. In three minutes the continuous hot strip mill turns a thick, six-foot-wide slab into a thin strip of sheet a quarter of a mile long. After pickling and cleaning, the cold reducing mill further flattens the steel into a ribbon one and a third miles long. The hot-strip mill passes the slab through a series of rollers, each rotating faster than the previous one because of the increasing length of the material. The strip is cut into single sheets or rolled into coils.

After cooling, the coils are either shipped as they are or processed to remove scale and surface oxidation and prepared for cold rolling. This is a process called "pickling" and consists of passing the steel through sulfuric acid, then into a tank for a cold-water rinse, a hot-water rinse, a dryer, and finally the steel is coiled again.

The cold-reduction process reduces the thickness of the hot rolled strip. Since drastic reduction in thickness hardens the steel, cold-reduced steel is heated to soften it; this process is called "annealing." Further treatment can also consist of coatings to protect against corrosion, to increase attractiveness of surface.

The finishing in widest use is the production of tin plate, which consists of steel coated with a thin layer of tin. In modern mills, tin plate is manufactured by electrolysis, which takes tin out of the electrolytic solution and deposits it on cold steel strip. This is done by passing the strip through a pickling solution and rotating brushes to clean it, then through a chemical solution where it is electroplated; a thin coat of oil is placed on the negatively charged plate, the strip is run through a cloud of positively charged plates of pig tin which is deposited on the negative steel. The steel strip coated with tin is then heated by high frequency coils; this causes the tin to melt and to flow to form a lustrous coat. By the time the molten tin coat is quenched in water, it is of uniform thickness.

Steel plates are rolled from slabs. They are used for such building as bridges, ships, rolling stock, storage tanks, and pressure vessels. The slabs are reheated slowly in reheating furnaces; they are then passed either through a mill which has only horizontal rolls and produces a plate with uneven edges and ends, called a sheared plate mill (products here have to be trimmed on all edges); or they are passed through universal mills, which have in addition to horizontal rolls at least a pair of vertical rolls, so placed as to roll horizontal edges of the plates straight and parallel.

The plate mill is usually of the two-high reversible type. The slab first passes back and forth as the rollers are drawn closer and closer together after each pass.

The hot-strip mill is used to produce sheets up to a quarter inch thick. The mill has roughing and finishing rolls mounted in tandem, each roll moving successively faster because of the elongation of the material. Sheets may receive various types of treatment, depending on their ultimate use. Generally, a relatively small plant (capacity less than half a million tons per year) is equipped with a reversing mill, while larger ones have a semicontinuous mill (about 1 million tons), and still larger plants (one and one-half million tons) have a continuous mill.

Auxiliary Facilities

Integrated (and even semi-integrated) steel mills have a number of auxiliary facilities necessary to keep the whole complex functioning smoothly. Each type of firm usually has extensive workshops both to repair and to produce parts for various sections. In Brazil and many other developing countries that have built steel mills, these shops are usually more extensive than in more advanced countries. The main reason is that the market for many parts is not large enough for specialized firms to be formed. Thus, in order to avoid shortages, the shops are generally larger than those to be found in United States or European steel firms. Actually, this has certain advantages. Workshops are labor intensive in nature and they contribute to the training of specialized labor.

All integrated firms usually have their own foundries. Some use these foundries to make individual iron or steel products for other firms. The activities of foundries in integrated steel firms, however, are devoted largely to cast ingot molds and stools to be used in the steel shop of those firms, as well as for making the most important cast spare parts.

Large integrated steel firms also have byproducts plants attached to the coke-oven section. We already mentioned above the type of products that may be produced. If an LD process is used in the steel shop, an oxygen-generating plant will be needed. In some cases the main workshop is used to repair the firm's rolling stock; sometimes special shops exist for this purpose. Many other sections may exist, such as, a water treatment plant or a department to service the electric power system.

TECHNOLOGICAL CHOICES

The economist concerned with the capital- as opposed to labor-intensive techniques will find in studying steel technology that few choices are open to him. This becomes obvious when examining each major section of a steel mill.

The blast furnaces of large firms, producing 1,000 or more tons of pig iron a day, cannot rely on labor-intensive methods of charging the furnace; given the large amounts of materials that have to be charged, the use of large and automatized methods becomes inevitable, for example, the use of huge stockhouse bins for the iron ore, coke, or limestone, the automatic mixing of these in a scale car, the dumping into a skip car which goes up an automatic skip hoist to drop the charge at the top of the blast furnace. Very modern installations even use conveyor belts instead of skips. The charcoal-burning furnaces in Brazil have a greater capacity for labor absorption, since the collection of wood in the countryside and its conversion into charcoal, and the systematic plantation of eucalyptus trees for such purpose is quite labor-intensive.[7] The making of coke is also a technologically rigid process; coke ovens cannot be adjusted in such a way as to make substantial capital savings by absorbing more labor.

The same situation holds for the steel-making section of a firm. Given the basic technology of an SM or LD furnace, there is little opportunity for factor substitution. It should be noted, however, that the degree of labor absorption is somewhat higher in the steel shop. The relining of the furnace with refractory brick is more frequent and is by its very nature labor intensive; there is also the constant need for relining of ladles used for pouring the liquid steel into ingots. (Probably the continuous casting system would be a substantial labor saving introduction. It bypasses the ingot preparation and casting stage in the steel shop and the initial process (blooming or slabbing) in the rolling mill. When the time comes to

---

7. It is doubtful whether in the long run, this type of labor-intensive activity is promising. The wood collection, plantation activities, and charcoal production depend on an abundance of very low-paid labor in the countryside.

consider this method for expanding the industry, some thought will have to be given about the profitability of the method chosen in terms of increasing the return to capital as opposed to reducing the labor force.)

Choices of production processes in the rolling mill are even smaller. In a larger plant, this is the most capital-intensive section. In older mills, semiautomatized equipment is often used, and in many smaller Brazilian mills one may still find workers pushing billets back and forth through rollers. This, however, is an inefficient method and cannot be used at all in the production of larger flat products.

A number of marginal operations absorb some labor—for example, scarfing; but even here machines are used when the scale of operation is very large. A substantial amount of specialized labor is absorbed by repair crews connected with rolling mills.

Beyond doubt, the most labor-intensive processes of a steel plant are found in its auxiliary sections, such as in the workshop and foundry. (Of course, the bureaucratic sections of Brazilian steel mills, such as office work are substantially more labor intensive than in the United States or Europe.) Although a foundry may be automatized to some extent, as in the preparation of the sand to be used in shaping the molds, most Brazilian foundries still use relatively labor-intensive processes.

Although the extent of choice of labor versus capital-intensive technology in a steel plant's marginal activities, such as the handling of materials is debatable, in the newer Brazilian firms relatively capital-intensive methods are used in unloading and stockpiling materials. To my query of why more labor-intensive methods were not adopted, at least for this section, engineers usually claimed that a large, modern steel mill consumed such large quantities of iron and coal that labor-intensive methods of unloading and stockpiling would be slow and expensive; they would create bottlenecks and result in a substantial rise in costs.

In sum, it may be stated with a fair degree of certainty that the economist's concern about labor- versus capital-intensive technology is largely irrelevant to the planner of a modern steel mill. As a matter of fact, it will be seen further on (Chapter Six) that labor cost as a proportion of value added is so small in most types of mills that the degree of labor use is not of basic concern to the planners of a steel operation. This does not imply, however, that a steel mill has little impact on employment. It will also be shown further on that in Brazil steel has had a considerable direct and indirect impact on the employment of the labor force.

In selecting the type of installation needed for a blast furnace, the main decision to be made consist of the size of the furnace—whether to build one large or two smaller furnaces or whether to include extensive ore beneficiation plants. The main measure of blast furnace efficiency is the "coke rate," that is, the quantity of coke used per ton of pig iron produced. The reason for this measure is that coke is the dominant cost item of blast furnace input, often amounting to 70 to 80 percent of the value of raw material input in Brazil.

The coke rate may be reduced by the use of agglomerated ore, especially the use of sinter (it may also be reduced by the injection of fuel and oxygen). The most relevant consideration concerning the decision of whether to install a beneficiating installation is its cost.

In a number of plants it might be considered worth having a higher coke rate for a while if there exists a shortage of capital for the construction of such an installation.[8] For the conditions prevailing in Brazil, however, it is more advisable to erect first the sinter plant, importing the necessary coke in this period, then install the coke oven and by-product plant, since the investment for the latter is much higher than for a sinter plant and the necessary ore-blending installation.

It is fairly safe to say that in the construction of new mills the steel shop will either consist of LD or electric furnaces. This has been confirmed by most competent sources. For example, a special United Nations inquiry into steel-making processes concluded that "oxygen converters of different types and electric-arc furnaces will be built almost exclusively in new steel shops."[9] The capital cost of an LD furnace is much lower than an open hearth (Siemens Martin) furnace; the capital savings can range from one third to more than one half (calculations made on the basis of capital charge per ton of steel), depending on the capacities compared.[10]

8. The capacity of blast furnaces is extremely flexible. It can be expanded not only by the use of beneficiated ores, which might not have been envisioned when the furnace was first erected, but also by the injection of fuel, high blast temperature, and high top pressure. Capacity can also be expanded when relining a furnace with thinner but better quality refractory. Many companies over the years have witnessed their furnaces producing 50 to 100 percent more than the originally estimated capacity.

9. United Nations, *Comparison of Steel-Making Processes*, p. 77.

10. See United Nations, *Interregional Symposium on the Application of Modern Technical Practices in the Iron and Steel Industry to Developing Countries,* Prague-Geneva (November 1963), p. 105; also see Walter Adams, and Joel B. Dirlam, "Big Steel, Invention, and Innovation," *The Quarterly Journal of Economics* (May 1966), pp. 179–180; the latter also describes the amazing slowness with which the large American steel firms adopted the LD method. See also G. S. Maddala and Peter T. Knight, "International Diffusion of Technical Change—A Case Study of the Oxygen Steel Making Process," *The Economic Journal* (September 1967).

Also, the operating cost of an LD furnace is about one third to one half lower than that of an SM furnace. Only the cost of the metallic charge per ton of steel produced is slightly higher in the former than in the latter. Calculations of total average costs have, however, given the edge to the LD system. The use of electric steel furnaces is generally more frequent in smaller specialized mills, where plenty of cheap electricity and scrap are available.

Where an SM system already exists, these furnaces can be expanded and made more productive by the injection of oxygen. Thus the facilities installed on the basis of the open-hearth system may be expanded for a while before new LD units are installed.

The capacity of the rolling-mill section of a steel plant is usually less flexible than the other sections. That is, the blast furnace section and the steel shop can be expanded more gradually than the rolling mill, which is more "lumpy" in nature. Rolling mills may be installed only in certain indivisible sizes. One therefore finds very often that in planning a steel mill which will expand in various stages, the rolling-mill section in the initial stages will have substantial amount of idle capacity, that is, it will be substantially underutilized. Some technological choices exist, however.

Once the final product has been decided on, the process to be adopted in the rolling mill depends on the production level the firm expects to reach. For example, one group of technicians has found that it would not pay to set up any type of hot-strip mill below a production level of 270,000 tons a year. Above that level, until about 500,000 tons a year, it pays to have a semicontinuous mill. A continuous mill pays only when the production level reaches 700,000 tons a year or more. It is therefore the size of the market that in this case determines the technology to be used, rather than relative factor costs or factor endowment.[11]

COST OF AN INTEGRATED STEEL MILL

It is extremely difficult to determine with much accuracy the cost of new integrated steel works. The initial cost estimates are usually prepared by a specialized engineering firm. These estimates are based on a variety of sources, some of which give exact information while others simply provide the engineers with an informed guess. Let us briefly outline the

11. M. D. J. Brisby, P. M. Worthington, and R. J. Anderson, "Economics of process selection in the iron and steel industry," *Journal of the Iron and Steel Institute* (September 1964).

procedure used in a developing country for planning investment in a new steel firm.

One of the first steps will be to determine the market for the steel products under consideration. This may be done in a developing country by taking into account the size of imports to be substituted then linking the growth rate of demand for steel products to some satisfying economic aggregates for projection purposes. Once the market, and thus the size of the desired steel mill, has been determined, the location of the plant will have to be decided upon. The traditional view about locational patterns is that a steel complex should be raw material–oriented, that it should be close to the most expensive raw material input—coal. Since in many European countries coal is imported, and this is the case in Brazil, it has been argued that a steel mill should ideally be located on the coast, and the pattern in recent years seems to be in that direction. In most developing countries, however, the location pattern is also heavily influenced by non-economic considerations, that is, political factors. Once these over-all location problems have been resolved and the general area of the plant determined, the engineering firm will try to find a place where there is an abundance of water, since the consumption of water is a main input of most sections of the mill. Also under consideration will be the availability of electric power and of an adequate transport system.

In developing countries, a transport system might be considered part of the basic steel investment decisions, that is, the development of a transport network is considered a complementary activity upon which the government might want to embark. Some attention also has to be given to the nature of the site chosen. Poor soil conditions, for example, can increase construction costs substantially and raise the cost of the entire project.

Let us briefly examine the main installations that will be necessary. If the blast furnace is to use coke, a coke plant will have to be planned. In all probability, a sintering unit will also be installed, since most new plants use some type of agglomerating process. The blast furnace, steel shop (with an adjoining oxygen plant, since most new plants are based on the LD process) and the rolling mill installations follow. The latter includes investment in the construction of soaking pits and a number of reheating furnaces placed at the beginning of each new rolling operation. Other major equipment will include unloading and loading facilities, extensive rolling stock, different types of cranes, and a large and complex quantity of electrical equipment. We have considered up to now only the equipment. A large proportion of expenditures goes, however, into construction of the various rather complex buildings.

Tables 1 to 6 should give the reader an idea of the relative magnitudes involved. They contain information derived from a number of feasibility studies prepared largely by foreign firms for a number of steel mills in Brazil: Usiminas, prepared in the mid-fifties and constructed in the latter fifties and early sixties; Cosipa, two plans prepared in 1957 and 1958, the Kaiser project being adopted for construction in the period 1959 to 1965; the expansion plans for Volta Redonda, which amount to the construction of an entirely new and parallel set of installations, since the capacity is to be substantially more than doubled—plans presented in 1965; and finally two projects prepared in 1962 and 1964 for the expansion of the firm Ferro e Aço Vitória, which at present operates only a rolling mill plant for nonflat products, into an integrated steel mill. The numbers presented in the tables represent the percentage distribution of expenditures between various units of the plant and also the division between materials purchased domestically and abroad. The tables are not really comparable. The estimates obtained contained information presented in different forms and in different degrees of detail. Domestic costs were usually converted into dollar costs at the free exchange rate prevailing at the time the studies were made. The lack of comparability does not really matter, since the main idea of the table is to give the reader relative magnitudes involved, and the different estimates contain different types of relevant information.

When examining the estimates in the first four tables, it is obvious that the rolling mill section absorbs most of the investment costs, from 30 to 44 percent of the fixed investment costs. It is also interesting to note that the cost of the coke plant is relatively equal to the weight of the blast furnace. This may be taken to imply that a charcoal-based steel mill would have substantial capital savings in the production of pig iron, so far as equipment is concerned. It will require, however, large investments in land purchases and reforestation. Once these investments are made, an appreciable period of time elapses before the production of this reducing agent is initiated. Also, the use of charcoal as a reducing agent eliminates the possibility of benefitting from economies of scale.

It is interesting to note the decrease in the import content of investment projects in Brazil in comparing the first three tables, representing projects planned in the period 1956–58, and the last three tables representing projects planned in the mid-sixties. Whereas the foreign cost of investments amounted to about 60 to 68 percent of total expenditures in the 1950s, it dropped to a range of 25 to 39 percent. This reflects the impact of the import-substitution industrialization which made it possible for Brazil to rely to a greater extent on its own sources of supply in planning new steel firms. For example, it will be noted in Table 3 that the foreign cost of the blast furnace section amounted to more than two thirds in planning Cosipa, while the plans for new blast furnaces of Volta Redonda in the late sixties need a foreign cost component of only 35 percent; the decline in the import content of the construction of the LD plant was from about 70 percent to about 33 percent; a sharp decline may also be witnessed in the foundry construction. The import-content decrease was less drastic in the rolling mills, where the basic machinery still had to be imported.

There now exists in Brazil, however, capacity for producing medium-sized rolling mills, for producing various types of cranes, and a number of other relevant capital goods.

Table 1

Usiminas: Percentage Distribution of Expenditures
on the Construction of Steel Works

Blast furnace and sintering plant	15.8%
Coke plant	8.7
Byproduct plant	3.6
Steel plant	6.0
Slabbing mill plant	9.5
Hot rolling mill	26.2
Cold rolling mill	8.6
Power plant	6.5
Transportation equipment	6.0
Electric power system	3.7
Service department	5.4
Total	100.0

Regional Distribution of Expenditures:

Japan	57%
Other Foreign	8
Brazilian	35
	100

Source: Unpublished report of the Banco Nacional de Desonvolvimento Econômico, 1956. These are expenditure plans for a 1.5 to 2.0 million ton steel plant. The original cost was estimated at US$ 238 million.

Actual cost after completion was estimated at 385 million dollars. See text for further explanation.

Table 2

Cosipa: Koppers Estimate of Construction Costs, Percentage Distribution[a]

	USA		Expenditures in Europe		Brazil				Total	
					Labor		Material			
Material handling	—	—	6.0	(55.5)	5.0	(22.0)	4.1	(22.5)	3.9	(100.0)
Sintering plant	—	—	4.0	(60.9)	2.8	(20.5)	2.0	(18.6)	2.4	(100.0)
Coke plant	—	—	14.6	(60.6)	12.1	(24.3)	6.2	(15.1)	8.7	(100.0)
Power station and distribution system	3.0	(8.8)	13.6	(57.6)	12.0	(24.6)	3.6	(9.0)	8.5	(100.0)
Blast furnace plant	—		12.4	(56.3)	11.6	(25.3)	6.9	(18.4)	8.0	(100.0)
Oxygen bessemer plant	—		11.1	(62.1)	6.6	(17.8)	6.1	(20.1)	6.5	(100.0)
Rolling mill plant	75.3	(52.4)	19.7	(20.1)	32.5	(16.0)	19.1	(11.5)	35.6	(100.0)
General facilities	—		13.3	(38.1)	17.4	(24.1)	22.3	(37.8)	12.6	(100.0)
Spare parts	4.8	(42.2)	4.5	(57.8)	—		—		2.8	(100.0)
Ocean freight	3.6	(17.3)	—		—		19.9	(82.7)	5.1	(100.0)
Consular expenses— insurance, etc.	1.3	(50.0)	0.8	(50.0)	—		—		0.7	(100.0)
Dock expenses and freight in Brazil	—		—		—		4.2	(100.0)	0.9	(100.0)
General engineering, construction, management	12.0	(71.4)	—		—		5.6	(28.6)	4.3	(100.0)
Total	100.0	(24.8)	100.0	(36.3)	100.0		100.0	(38.9)	100.0	(100.0)

Source: Computed from Companhia Siderúrgica Paulista, "Cosipa," *Report for Program for Construction of Steel Works at Piaçaguera for Production of Flat Rolled Products*, Volume IV, Economics, Organization and Finance, February 1958, Foreign Department, Koppers Company, Pittsburg, Pa., unpublished.

Note: The original estimated total cost of Koppers was US$ 168 million.

[a] Vertical percentage distribution—nonparenthesized numbers, reading downward, give proportion of total expenditure of each section in steel mill. Horizontal percentage distribution—numbers in parenthesis, reading across, numbers in parenthesis give proportion of total expenditure in each section abroad and in Brazil.

Table 3

Cosipa: Kaiser Estimate of Construction Costs, Percentage Distribution[a]

	Purchases in U.S.			Brazilian Costs				Total
	Purchase of machinery and equipment	Purchase of permanent materials	Ocean freight and marine insurance	Purchase of machinery and equipment	Purchase of permanent materials	Freight and dock handling	Installation	
Site preparation and yard development	4.2 (28.6)	— —	3.7 (2.5)	19.5 (9.2)	12.9 (24.9)	8.7 (0.4)	17.9 (34.4)	8.4 (100)
Raw-material handling	1.1 (22.1)	— —	1.0 (2.0)	34.2 (47.8)	3.4 (19.8)	2.3 (0.3)	6.6 (7.8)	2.8 (100)
Boiler and power plant	7.7 (72.3)	— —	7.4 (6.9)	— —	6.4 (17.2)	1.2 (—)	3.9 (3.6)	6.1 (100)
Blast furnace	4.4 (37.3)	36.3 (22.8)	8.5 (7.3)	2.1 (1.2)	6.8 (16.5)	12.7 (0.6)	6.8 (14.3)	6.7 (100)
Sinter plant	1.3 (50.3)	2.0 (5.6)	1.6 (6.3)	2.3 (6.0)	1.4 (13.9)	4.2 (1.0)	1.7 (16.9)	1.5 (100)
LD plant	4.4 (52.4)	11.7 (10.1)	5.9 (6.9)	5.8 (4.2)	2.8 (9.4)	4.8 (0.5)	4.6 (16.5)	4.9 (100)
Oxygen generation plant	1.8 (74.3)	0.4 (1.2)	1.7 (7.2)	2.0 (5.8)	0.6 (6.3)	0.8 (0.3)	0.7 (4.9)	1.4 (100)
Coke ovens	3.0 (35.8)	9.4 (8.2)	3.4 (4.1)	2.2 (1.8)	12.6 (42.7)	4.4 (0.3)	7.7 (7.1)	4.8 (100)
Byproduct plant	4.2 (74.2)	— —	4.2 (8.6)	— —	2.1 (10.9)	0.6 —	1.7 (6.3)	3.2 (100)
Ingot stripper	0.7 (74.8)	0.5 (3.4)	0.8 (8.2)	0.7 (4.5)	— (1.5)	0.4 (0.2)	0.2 (7.5)	0.6 (100)
Soaking pits	3.2 (51.9)	3.5 (4.2)	3.7 (5.9)	1.4 (1.5)	4.4 (20.1)	5.6 (0.5)	4.2 (15.9)	3.5 (100)
Rolling and finishing facilities								
Buildings	— —	27.5 (14.5)	4.2 (3.0)	5.8 (2.8)	17.3 (35.2)	31.8 (1.4)	20.4 (43.0)	8.0 (100)
Mechanical equipment	27.5 (85.0)	1.2 (0.3)	27.9 (8.6)	— —	2.7 (2.4)	15.6 (0.2)	3.6 (3.5)	18.7 (100)
Electrical equipment	23.2 (81.5)	— —	11.7 (4.1)	— —	13.0 (12.9)	0.9 (—)	7.3 (1.5)	16.4 (100)
Other equipment	9.8 (69.5)	4.2 (2.2)	10.2 (7.2)	20.3 (9.9)	3.6 (7.3)	5.2 (0.3)	5.3 (3.6)	8.1 (100)
Ingot mold foundry	1.2 (53.0)	2.4 (7.8)	1.5 (6.9)	1.5 (4.6)	0.7 (7.4)	0.6 (0.2)	1.3 (20.1)	1.3 (100)
Mold preparation	— (14.1)	0.9 (17.5)	0.3 (5.0)	1.4 (23.5)	— (12.5)	0.2 (0.3)	0.6 (27.1)	0.3 (100)
Buildings and shops	2.3 (41.6)	— —	2.3 (4.2)	0.8 (1.0)	9.3 (46.3)	— (—)	5.5 (6.9)	3.3 (100)
Total	100.0 (57.7)	100.0 (4.2)	100.00 (5.7)	100.0 (3.9)	100.00 (16.4)	100.0 (0.4)	100.0 (11.7)	100.0 (100)

Source: Computed from Kaiser Engineers International Inc., *Engineering Report on Cosipa Steel Mill Project for Cia. Siderúrgica Paulista,* at Piaçaguera, Brazil, August 20, 1957.

Note: The original estimated total cost was approximately US$ 151.5 million.

[a] Vertical percentage distribution—non-parenthesized numbers.

Horizontal percentage distribution—numbers in parentheses.

The Brazilian Steel Industry

Table 4

Volta Redonda: Estimate of Construction Cost of Plant Expansion
from 1.7 to 3.5 Million Tons, Percentage Distribution[a]

	Foreign Costs	Brazilian Costs	Total
Mining and Beneficiation	0.3 (14.0)	0.8(86.0)	0.6(100.0)
Agglomeration	5.6 (35.8)	6.4(62.4)	6.1(100.0)
Handling of raw materials	0.7 (26.9)	1.3(73.1)	1.1(100.0)
Coke Plant and Byproducts	6.7 (26.0)	12.1(74.0)	10.0(100.0)
Modernization of blast furnace 2	0.3 (15.1)	0.8(84.9)	0.6(100.0)
Building of BF 3 and 4	6.9 (35.7)	7.9(64.3)	7.5(100.0)
Modernization of SM furnace	0.02(5.7)	0.2(94.3)	0.1(100.0)
New LD plant	5.4 (33.1)	6.8(66.9)	6.3(100.0)
Slabbing mill	11.7 (52.1)	6.7(47.9)	8.6(100.0)
Hot strip mill	17.3 (48.2)	11.7(51.8)	13.9(100.0)
Hot finishing mill	0.2 (11.4)	0.5(88.6)	0.4(100.0)
Cold strip mill	4.2 (48.9)	2.8(51.1)	3.3(100.0)
Cold finishing line mill	1.0 (4.1)	0.4(95.9)	0.6(100.0)
Galvanizing installations	1.4 (41.0)	1.2(59.0)	1.3(100.0)
Electrolytic tinning installations	11.6 (51.6)	6.9(48.4)	8.7(100.0)
Rail and structural mill	0.6 (23.2)	1.0(76.8)	0.7(100.0)
Structural welding plant	1.0 (35.0)	1.3(65.0)	1.2(100.0)
Steam production installations	2.3 (38.8)	2.3(61.2)	2.3(100.0)
Thermoelectric station	1.1 (56.9)	0.5(43.1)	0.7(100.0)
Electric energy distribution	1.0 (40.6)	0.9(59.4)	1.0(100.0)
Combustible distribution system	0.4 (10.7)	2.0(89.3)	1.4(100.0)
Water pumping system	0.3 (19.2)	0.8(80.8)	0.6(100.0)
Oxygen plant	3.2 (54.3)	1.8(65.7)	2.3(100.0)
Rail lines and roads	— —	2.0(100.0)	1.2(100.0)
Rolling stock (internal)	3.0 (52.6)	1.1(47.4)	1.8(100.0)
Foundry	0.9 (13.5)	3.7(86.5)	2.6(100.0)
Maintenance shop and supply depot	0.6 (9.7)	3.8(90.3)	2.6(100.0)
General services	— (8.6)	0.6(91.4)	0.4(100.0)
Engineering costs	2.6 (43.9)	1.4(56.1)	1.9(100.0)
Miscellaneous	4.2 (40.8)	3.8(59.2)	4.0(100.0)
Spare parts shop	5.5 (82.1)	0.8(17.9)	2.6(100.0)
Metallic iron alloy plant	— —	0.4(100.0)	0.3(100.0)
Railroad cars	— —	5.3(100.0)	3.3(100.0)
Total	100.0 (38.7)	100.0 (61.3)	100.0 (100.0)

Source: Computed from data in: Companhia Siderúrgica Nacional, *Expansão De Volta Redonda, Plano D,* Rio de Janeiro, 1965. Computed from dollar values.
  [a] Vertical percentage distribution—nonparenthesized numbers. Horizontal percentage distribution—numbers in parentheses.

Table 5

Companhia Ferro e Aço Vitória: Estimated Expansion Cost According to McKee—From Simple Rolling Mill to Integrated Steel Works: Percentage Distribution of Construction Costs[a]

	Brazilian			
	Material	Labor	Foreign	Total
Coke ovens	15.2 (37.4)	17.0 (34.6)	10.9 (28.0)	14.2 (100.0)
Blast furnace	15.0 (37.0)	14.8 (30.4)	12.7 (32.6)	14.2 (100.0)
Steel plant	14.2 (34.3)	13.7 (27.3)	15.2 (38.4)	14.4 (100.0)
Power plant	6.1 (15.7)	13.8 (29.2)	20.6 (55.1)	13.6 (100.0)
Oxygen plant	1.6 (20.9)	2.0 (21.0)	4.3 (58.1)	2.7 (100.0)
Roads and railroads	7.1 (64.9)	4.6 (34.7)	— (0.4)	3.8 (100.0)
Rolling stock	1.7 (29.5)	— (1.3)	3.9 (69.2)	2.1 (100.0)
Utilities	4.4 (31.5)	4.1 (24.0)	6.0 (44.5)	4.9 (100.0)
Office, shops	3.3 (47.4)	3.3 (40.0)	0.8 (12.6)	2.3 (100.0)
Rolling mill[b]	15.0 (63.2)	5.7 (19.7)	3.9 (17.1)	8.3 (100.0)
Other expenses (engineering, freight, etc.)	6.4 (17.2)	13.3 (29.5)	18.9 (53.3)	12.9 (100.0)
External to plant (water pumps, power transmission lines, etc.)	10.1 (52.4)	7.6 (32.8)	2.8 (14.8)	6.6 (100.0)
Total	100.0 (34.9)	100.0 (28.7)	100.0 (36.4)	100.0 (100.0)

Source: Report Prepared for the Companhia Ferro e Aço de Vitória, Brazil "Expansion and Integration of Operations," June 1964, prepared by Arthur G. McKee and Company Engineers and Contractors, Cleveland, Ohio.

Estimated cost of expansion–US$ 91.8 million, for capacity of 380,000 tons/year of products.

[a] Rolling mill consists of upgrading of already existing plant-blooming mill, structural mill and bar and rod mill.

[b] Vertical percentage distribution–non-parenthesized numbers. Horizontal percentage distribution–numbers in parentheses.

Table 6

Companhia Ferro e Aço Vitória: Construction Cost Distribution for Alternative Plant Sites and Various Construction Periods—Falcão Estimates—Percentage Distribution[a]

	Capuaba Plant (700,000 tons cap.) 7 year Construction			Camburi Plant (4 years)			Laranja Plant 5 year Construction		
	Brazilian	Foreign	Total	Brazilian	Foreign	Total	Brazilian	Foreign	Total
Material and equipment	20.1 (32.6)	82.6 (67.4)	40.9 (100.0)	26.3 (69.8)	72.2 (30.2)	32.6 (100.0)	24.6 (50.8)	74.3 (49.2)	36.7 (100.0)
Construction and installation	45.2 (100.0)	—	30.1 (100.0)	35.6 (100.0)	—	30.8 (100.0)	37.1 (100.0)	—	28.0 (100.0)
Engineering and administration	5.6 (70.9)	4.6 (29.1)	5.3 (100.0)	5.8 (69.0)	16.7 (31.0)	7.3 (100.0)	5.5 (69.0)	7.7 (31.0)	6.0 (100.0)
Construction equipment	1.9 (22.9)	12.8 (77.1)	5.5 (100.0)	0.6 (25.0)	11.1 (75.0)	2.0 (100.0)	1.9 (25.0)	18.0 (75.0)	5.8 (100.0)
General construction expenditures	4.3 (100.0)	—	2.9 (100.0)	4.1 (100.0)	—	3.5 (100.0)	4.3 (100.0)	—	3.3 (100.0)
Provisional construction	1.4 (100.0)	—	0.9 (100.0)	1.2 (100.0)	—	1.0 (100.0)	1.4 (100.0)	—	1.0 (100.0)
Construction headquarters	3.2 (100.0)	—	2.1 (100.0)	2.9 (100.0)	—	2.5 (100.0)	3.2 (100.0)	—	2.4 (100.0)
Technical organization and experimental operations	9.2 (100.0)	—	6.2 (100.0)	11.8 (100.0)	—	10.2 (100.0)	11.0 (100.0)	—	8.4 (100.0)
Residential area	9.1 (100.0)	—	6.1 (100.0)	11.7 (100.0)	—	10.1 (100.0)	11.0 (100.0)	—	8.4 (100.0)
Total	100.0 (66.6)	100.0 (33.4)	100.0 (100.0)	100.0 (73.4)	100.0 (26.6)	100.0 (100.0)	100.0 (75.7)	100.0 (24.3)	100.0 (100.0)

Source: Companhia Ferro e Aço Vitória, Estudo Para Escolha Do Local Da Etapa De Integração, Rio de Janeiro, 1962. Study Made by Engineer José Mariano Falcão.

Total Estimated US$ costs:

Capuaba	US$ 169 million
Laranja	128
Camburi	106

[a] Vertical percentage distribution—nonparenthesized numbers.
Horizontal percentage distribution—numbers in parentheses.

Some information, although not very detailed, is available to complement the data of these tables. First of all, since Usiminas was located in an uninhabited area, the company had to spend substantially to build a city for its employees. This expenditure is not included in the Table 1 estimates. Also, a construction period longer than scheduled and the inflationary situation in Brazil raised capital costs considerably above estimates. Thus, instead of the originally estimated cost of 238 million dollars (excluding the city), actual costs were about 385 million (this includes about 60 million dollars for the construction of a city). Although the Kaiser project, which was accepted, estimated a cost of 151.5 million dollars for the Cosipa project (this excludes financial costs), the estimate before construction began was for about US$ 216 million which includes an additional US$ 50 million for financial expenditures, such as, interest on debt. The actual cost of Cosipa was about US$ 299 million, excluding financial expenditures. This substantially higher actual cost was mainly a result of the very high construction costs for the mill. It was discovered only after work was begun near Santos that soil conditions were extremely bad, thus raising construction costs from an estimated US$ 103 million to an estimated US$ 190 million. Unfortunately, there is no breakdown available for actual expenditures. Except for the Cosipa construction difficulties, however, the proportions in the tables are an accurate representation of the cost distribution.

It would be hazardous to try to extrude from this information a capital-output ratio for the steel industry or an accelerator type of relationship. Although Usiminas and Cosipa respectively were producing at an annual rate of about 600,000 and 400,000 ingot tons in 1967, the rolling mill capacities of each of these plants can be estimated conservatively at 1.5 million tons. The 1966 expansion plans of Usiminas and Cosipa, costing US$ 80 and US$ 50–60 million respectively, will increase their actual productive capacity to a million tons each. The sum of the basic costs and the latter output, however, do not give an accurate picture of the capital-output ratio. Most newly constructed integrated steel mills will deliberately construct rolling capacity substantially greater than pig iron and steel-producing capacity. The main reason for this is that the rolling mill represents the largest and most expensive part of the investment. Also, the nature of a rolling mill is such that it is far more expensive, if not impossible, to build it up gradually. For example, in order to produce wide plates of more than 2½ meters, a mill of about one-million-ton capacity is necessary, even though at the time of installation of such a mill there is no market to use such a plant fully. On the other hand, the blast furnace and steel-shop sections can more easily be built up gradually.

Taking into account the lumpiness of investment in many sections of a steel plant and given the cost of new facilities in Brazil in the fifties and sixties compared with the value of the marginal output achieved, the marginal capital-output ratio stood at approximately 3 to 5. The ratio is expected to be smaller in the future (late 1960s and 1970s) because in order to expand rolled-steel products, only the blast furnace and steel-shop sections will undergo expansion in some of the most important firms (especially Cosipa and Usiminas). Some engineers have estimated that the marginal capital-output ratio might fall to about 2.8 in the decade 1965–75 for the steel industry as a whole.

3

AVAILABILITY OF
STEEL-MAKING INPUTS IN BRAZIL

WITH THE exception of good quality coal for coking, Brazil possesses most of the natural resources required for steel production. Let us examine briefly the location and reserves of the principal resources.

IRON ORE

Brazil has one of the world's largest iron ore reserves (it ranks third in the world) and the largest reserves to be found in Latin America. According to estimates made by a group of the Economic Commission for Latin America, the known reserves in 1962 were greater than 15 billion tons, or about three fourths of the reserves of Latin America. The potential resources—ores with a relatively lower yield—amounted to an estimated 23 billion tons, an even greater proportion of the total Latin American potential (see Table 7). Indeed, one later estimate claims that Brazil's actual reserves amount to 30 billion tons, and a most recent joint survey of the Divisão de Fomentação de Produção Mineral and the United States Geological Survey estimate reserves of 4.5 billion tons of hematite and 23.5 billion tons of itabirite. The iron content of the primary reserves (composed mainly of hematite) of the country vary between 58 and 66 percent.[1] This is high compared with the iron content of a number of other

1. Comissão Econômica Para a America Latina, *A Economia Siderúrgica da American Latina: Monografia do Brasil* (by J. M. Falcão), p. V-7. The estimate of world iron ore reserves in 1962 (according to the British Iron and Steel Federation) was 250 billion tons, of which the United States had 5.4 billions, Canada 11.0, Venezuela 2.2, Chile 3.0, France 4.4, United Kingdom 2.6, Sweden 3.9, Germany 5.5, the Soviet Union 77.0, India 21.0. The iron content of various countries were,

of the large iron ore producers. The secondary reserves consist primarily of itabirite (hematite and quartz) with 35 to 60 percent iron content. These have remained for the most part relatively untouched, but they have great potential applicability through beneficiation and pelletizing, especially for export purposes.

Compared with its reserves, Brazil's production of iron ore has been small, though the output for both steel production and iron ore exports has grown through the late 1950s and into the 1960s. Until 1937, the country never produced more than half a million tons annually. In the post–World War II period, total output increased substantially, the rate increasing very rapidly in the later fifties. From 1954 to 1964, production increased more than five times, spurred on both by an energetic export drive and increased domestic consumption. Even by 1965, however, total iron ore output of close to 20 million tons amounted to less than 0.1 percent of known reserves (not taking into account the potential reserves of smaller yields).[2] By 1966, Brazil ranked seventh in world iron ore production for domestic consumption and exports (after Russia, the United States, France, Canada, Sweden, and Venezuela). The bulk of Brazil's iron ore reserves is found in six states, though up to the mid-sixties about 99 percent of actual iron ore production originated in the state of Minas Gerais. Most of this production takes place in the center of the state, in an area of about 7,000 square kilometers, known as the "Quadrilátero Ferrífero."

Within Minas Gerais are two principal independent iron ore–producing zones. One is the zone of the Vale do Rio Doce, whose production is brought to the port of Vitória on the Vitória-Minas railroad. The largest portion of the country's iron ore exports go through this port. The second major iron-producing zone is the one in the region of the Paraopeba river and the valley of the Rio das Velhas, the zone being generally referred to as the Vale do Paraopeba. The latter's exports are shipped through the port of Rio de Janeiro. In 1965, the Vale do Rio Doce accounted for a production of about fifteen million tons and the Vale do Paraopeba five million. Table 8b gives the location and estimates of the reserves of the principal mines in Minas Gerais. It should be noted that some are owned

Sweden 64 percent; Venezuela 60 percent; United States 45 percent; Soviet Union 45 percent; France 38 percent; Germany 32 percent and United Kingdom 28 percent.

2. The estimates of iron ore reserves in Brazil in Tables 7 and 8, and those mentioned in the text, are not consistent. They represent estimates at various points of time.

Table 7

Latin America: Iron Ore Reserves and Potential Resources

(Millions of Metric Tons)

Country	Ore reserves	Yield iron (percent)	Potential resources	Yield (percent)	Total
Argentina	251[a]	48	500	4[b]	751
Bolivia	—	—	540	60	540
Brazil	15,421	58	23,059	35	43,480
Central Am.	20	60	—	—	20
Chile	337	62	1,250	51	1,587
Colombia	172[a]	48	380	325	552
Cuba	—	—	2,500	45	2,500
Dominican Rep.	50	67	—	—	50
Ecuador	1	66	—	—	1
Mexico	574	62	—	—	574
Peru	767	60	—	—	767
Uruguay	28	40	72	48	100
Venezuela	2,908	62	1,223	45	4,133
Total	20,529	60	24,526	40	55,055

Source: United Nations, *Interregional Symposium on The Application of Modern Technical Practices in the Iron and Steel Industry to Developing Countries*, Discussion Paper/ECLA.1, 22 October 1963.

[a] Phosphoric ores with 0.8 to 1.0 percent phosphorous
[b] Ferriferous sands.

by the steel companies themselves and some by firms engaged mainly in exporting iron ore, especially the Companhia Vale do Rio Doce. The latter, a government-owned corporation, is Brazil's largest mining enterprise. This company owns and operates the Vitória-a-Minas railroad, a one-meter-gauge line of 510 kms. which connects the Itabira mining area with the ports of Vitória and the new marine terminal at Tubarão, which can receive big bulk carrier up to 100,000 tons. The company's exports reached more than ten million tons in 1966. It is now erecting a pelletizing plant in Tubarão for export purposes to make use of its great reserves of ore fines.

Brazil's iron ore–export capacity will also be expanded as a result of the plans of the Mineração Brasileira Reunidas, owned by the Antunes group and a minority participation of American interests. This group intends to erect a pelletizing plant at the port of Tubarão or at the new terminal to be built in the Sepetiba bay area.

The rapid growth of Brazilian, and generally Latin American, exports

Table 8

a. Brazil: Iron Ore Resources and Reserves

(in millions of metric tons)

Location	Type[b]	Total Reserves[a]	Potential
Minas Gerais:			
320 km. northwest of Rio de Janeiro	Haematite (66)	2,096	
Minas Gerais:			
320 km. northwest of Rio de Janeiro	Haematite (50)	3,000	
Minas Gerais:			
320 km. northwest of Rio de Janeiro	Itabirite (35)	10,000	18,000
Urucum (Bolivian Frontier)	Itabirite (55)	300	10,000
Amapa (Mouth of Amazon)	Itabirite (66)	—	9
Bahia	Itabirite (60)	—	50
São Paulo	Magnetite (60)	3	—
Paraná	Magnetite (60)	22	—
Total		15,421	28,059

Source: Same as Table 7.

[a] Includes proven, indicated and calculated.
[b] Number in parentheses: iron content percentage.
Estimates for 1961.

b. Location, Reserves and Ownership of Brazil's Most Important Functioning Iron Ore Mines

Mine	Location	Estimated reserves (millions of tons)	Owner[a]	Type of Ore
Cauê	Itabira, M. G.	113	CVRD	Compact Haematite
Casa da Pedra	Congonhas do Campo, M. G.	158	CSN	Compact Haematite
Pico de Itabira	Pico de Itabirite, M. G.	30	MBR	Compact Haem. & Itabirite
Águas Claras	Belo Horizonte, M. G.	330		
Alegria	Caraça, M. G.	300		
Andrade	Monlevade, M. G.	30/40	CSBM	Haematite
Candu	R. Paraopeba, M. G.		MGB	Itabirite

Source: United Nations, Cepal, "A Economia Siderúrgica Da América Latina: Monografia do Brasil," mimeographed, prepared by J. M. Falcão, December 1964.

[a] CVRD—Companhia Vale do Rio Doce, government-owned iron ore exporting company.
CSN —Companhia Siderúrgica Nacional.
CSBM—Companhia Siderúrgica Belgo Mineira.
MGB —Mineração Geral do Brasil
MBR —Mineração Brasileira Reunidas.

Table 9

Brazil: Iron Ore Production, Exports and Export Prices

(Thousands of tons and US$/t)

Year	Iron ore production total	Iron ore production of Minas Gerais	Iron ore exports	Iron ore export prices Average
1944	770	764	206	—
1950	1,987	1,972	890	7.92
1954	3,071	3,051	1,678	12.86
1955	3,382	3,346	2,565	11.68
1956	4,075	4,044	2,745	12.80
1957	4,977	4,950	3,550	13.55
1958	5,185	5,152	2,831	13.93
1959	8,907	8,891	3,968	10.96
1960	9,345	9,242	5,240	10.24
1961	10,221	10,130	6,237	9.57
1962	10,737	10,691	7,528	9.08
1963	11,219	11,183	8,207	8.57
1964	16,960	16,924	9,730	9.08
1965			12,386	

Source: Conselho Nacional de Estatística do I.B.G.E.; Instituto Brasileiro de Siderurgia; United Nations, *A Economia Siderúrgica Da America Latina, Monografia Do Brasil,* prepared by J. M. Falcão.

Table 10

Latin America: Exploitation of Iron Ore Resources And Use of the Ore in 1962

(millions of metric tons)

Country	Production	Consumption	Exports Abroad or Within the Region
Argentina	0.13	0.65	− 0.51
Brazil	10.50[a]	3.00	+ 7.50[a]
Chile	8.09	0.60	+ 7.49
Colombia	0.42	0.42	—
Mexico	1.30	1.30	—
Peru	5.23	0.18	+ 5.05
Venezuela	13.20	0.80	+12.60
Total	38.87	6.95	+32.13

Source: Same as Table 7.

[a] A comparison with Table 9 will show this estimate to be slightly lower to the actual Brazilian output. However, this number was left intact here in order for the original comparisons made to hold.

of iron ore over the decade 1955 to 1965 resulted from the high iron content of the ores from the region and the low impurities such as sulphur and phosphorus they contained. This made Latin American iron ores (lump—2″ to 8″) an excellent material as an addition to the metallic charge of the open-hearth furnaces for cooling the bath "assisting in the refining process through the liberation of oxygen in order to oxydize some of the impurities contained in the molten bath. Technological developments of the latest years, particularly the adoption of the BOF process, and the use of oxygen in the open-hearth furnaces, are changing these conditions to the disadvantage of Latin American exporters."[3] As we discussed in the previous chapter, the tendency in the last decade has been to make increasing use of beneficiated ores. This has led Brazil and some other Latin American iron ore producers to plan for the establishment of installation for exporting beneficiated ores in the form of pellets.

Iron ore exports, which had risen from 2.5 million tons in 1955 to 12.5 million tons in 1965, are supposed to reach twenty-five to thirty million tons in 1970. Domestic consumption had reached about five million tons in 1966.

MANGANESE

Brazil possesses 80 percent of the known reserves of manganese in Latin America.

Table 11 gives the geographical distribution of known reserves of manganese ore in Brazil. Until the late fifties, Minas Gerais was the main producer of maganese; it still is, at this writing, the principal supplier of this ore to the domestic steel industry. The maganese content of the Minas Gerais ores vary between 36 and 41 percent, though in some areas the content has reached 53 percent (45 percent is considered to be a reasonable level for exports in the world market).

Table 11
Brazilian Manganese Reserves
(in millions of tons)

Amapá	31.0
Amazonas	0.3
Bahia	1.5
Espírito Santo	0.15
Minas Gerais	16.4
Mato Grosso	103.0
Other States	0.2
TOTAL	152.55

Source: Falcão-UN/ECLA study, *op. cit.* p. V–13.

3. United Nations, *Interregional Symposium.* . . . *op. cit.,* "Steelmaking Raw Materials in Latin America," p. 12.

The manganese of the Amapá territory (manganese content of ore about 46 percent) has been developed in the decade 1955–65 by ICOMI, a consortium of the Brazilian firm CAEMI and Bethlehem Steel, exclusively for export. Production there increased from 60 tons in 1956 to more than 900,000 tons in 1963. Huge deposits were discovered recently in the state of Mato Grosso, but the ore is not quite as rich as that of Amapá, and its location allows for transport only through small barges at higher transportation costs.

COAL RESERVES

The only coal deposits in Brazil are found at present in the southern part of the country, mainly in the state of Santa Catarina, although some isolated deposits can be found in the states of São Paulo, Paraná, and Rio Grande do Sul. The total reserves of the state of Santa Catarina have been estimated at 1.7 billion tons. Taking the other states into account, total Brazilian reserves have been estimated at between 2.1 and 3 billion tons. Most of the coal is bituminous, and only the Santa Catarina coal is usable for coking purposes. Even the latter is of relatively poor quality, however, because of its high ash content. The ash content of Santa Catarina coal amounts to 17.9 percent, compared with 4.1 to 3.8 for American coal of high and low volatility respectively. Studies made at the Volta Redonda plant have shown that the acceptable limit of ash content in coke is 13 percent; a higher content would severely hamper the effective functioning of the blast furnace. Given the above-mentioned ash contents of Brazilian and American coal, it was found that no more than 40 percent of Santa Catarina coal could be allowed in the mixture for producing coke.[4]

The annual production of coal in Santa Catarina in the mid-1960s amounted to about 1.5 million tons. Of this amount, after extensive washing operations, 47 percent was used as coking coal, 28 percent was used for steam-generating, and the other 25 percent was not found usable.[5] The high ash content necessitates substantial facilities for washing the coal. The Companhia Siderúrgica Nacional, which owns a number of coal mines in Santa Catarina, has a large washing plant in Capivari that can treat about 400 tons of coal per hour.

4. Amaro Lanari Junior, "Consumo De Carvão Nacional Na Siderurgia," *Metalurgia*—ABM, No. 93, XXI (1965), 647.
5. Ministério do Planejamento e Coordenação Econômica, EPEA, Siderurgia, *Metais Não-Ferrosos: Diagnostico Preliminar* (Abril 1966), p. 61.

The Companhia Siderúrgica Nacional is the major producer of coking coal. It's mines in Santa Catarina account for about 35 percent of total output. It has a strip mine in Siderópolis and pit mines at Criciuma. The latter is operated by the Prospera Company, a wholly-owned subsidiary. Four other large producers account for 50 percent of total production and six smaller ones for 15 percent. The CSN's coal-washing facilities at Capivari serve all firms.

With CSN, Usiminas and COSIPA using 40 percent domestic coal in producing coke, total consumption of this coal in 1965 amounted to 475,000 tons and estimates for 1966 are of 600,000 tons.

The difference in costs of imported versus national coal is large. One authority reported that in 1965 the cost of a ton of imported coal at Ipatinga, Minas Gerais (the location of Usiminas), was US$ 22.05, whereas the cost of domestic coal was US$ 42.45.[6] The high cost of domestic metallurgical coal is caused, in part, by its poor quality and the necessary washing operations. Most of the high cost, however, results from the inefficient operation of some mines, a deficient transportation system for coal, and the price policies followed by the Council of the Commission of the Plan for National Coal Comissão do Plano de Carvão Nacional). I shall comment on the effects of using national coal on the productivity of the blast furnaces in Chapter Six. The relatively low cost of American coal results partly from the fact that it is mined on a very large scale with modern equipment.

Until 1966, the attitude of the government was rather inflexible on the use of 40 percent national coal in producing coke. The principal reason for this attitude was the concern over the social effects of closing the Santa Catarina coal operations, especially the resulting unemployment. In 1966, the government adopted a less rigid position. The percentage of domestic coal use would become flexible. The 1966 coal-production level would be maintained, and any increase in output would be allowed only for alternative uses of the coal. In the case of steel production destined for exports, domestic coal requirements would be lowered in proportion to the ratio of exports to total production, assuming the 40 percent national coal participation as the basing point. Thus, a company which would export 50 percent of its production would have to consume only 20 percent of domestic coal.

6. Amaro Lanari Junior, "Consumo de Carvão Nacional na Siderurgia," *Geologia e Metalurgia,* No. 27 (1965), p. 221

CHARCOAL

As already mentioned in the previous chapter, Brazil is the world's largest user of charcoal blast furnaces. About 30 percent of the pig iron produced in Brazil is based on charcoal. In 1964, approximately 3.1 million cubic meters of charcoal were consumed by charcoal-based blast furnaces. Initially, all the charcoal used came from the natural forests in Minas Gerais. In the last decade, however, increasing use has been made of eucalyptus plantations.

The Belgo-Mineira company was the first and is still the largest integrated Brazilian steel producer using charcoal as its principal reducing agent in the blast furnace. The second largest charcoal consumer is the Acesita company. Other smaller integrated operations using charcoal are Barra Mansa, Aliperti, the CBUM organization and about 90 small, independently owned blast furnaces in the state of Minas Gerais.

The economic advantages of using charcoal-based blast furnaces in the past were the proximity of natural and/or planted forests, low transport cost of charcoal (since it is so close), the availability of cheap agricultural labor, and low land values. Of course, conditions have changed considerably. With the exception of Belgo Mineira and Acesita, which have large forested areas in the vicinity of their respective plants, the consumers of this reducing agent have to haul it over long distances. Charcoal production in Minas Gerais resulted in serious depletion of natural forests in the mining areas, which gave rise to legislation forcing producers to engage in reforestation on a one-to-one basis for cuts of natural trees. Reforestation is usually done with eucalyptus trees. The latter can be cut for the first time after 8 years, and second and third cuttings are spaced about 7 years apart. Investment costs in reforestation are relatively high for the small producer, making its application for iron and steel production less and less attractive.

The expansion plans of the Companhia Siderúrgica Belgo Mineira and of Acesita are based on the use of Charcoal. Belgo Mineira owns large areas of land covered by natural forests and eucalyptus plantations, totaling about 309,000 acres. The company originally used only natural forests. During the last ten years this company has been planting an average of 10 million trees per year, and in 1965, 13.5 percent of the charcoal produced was made from planted trees. This percentage is supposed to increase steadily over the next twenty years.

It is obvious, however, that continued growth of the charcoal consumption by the steel industry is feasible only if the cost of agricultural labor remains relatively low, if the land values do not increase to an extent that would make it uneconomic to use such land for growing wood for charcoal, if there is no increase in the value of wood for paper, plywood, and plastics production. Also, a considerable reduction in the delivered cost of domestic and imported coal could induce a substantial decline in the use of charcoal. Belgo-Mineira and some other firms are prepared to shift to the use of coke as soon as the advantages of charcoal-based pig iron production disappear.

FLUXES

Limestone deposits are abundant throughout Brazil, and most of the large mills of Brazil get their limestone from Minas Gerais. The distance it has to be transported varies between 200 and 300 kms. Brazil also has large deposits of high-grade dolomite in the states of Rio de Janeiro and Minas Gerais.

SCRAP

It will be remembered from the previous chapter that scrap is an important input in an SM steel furnace, that it is the principal input of an electric steel furnace, and that it is of much smaller importance as an input of an LD furnace. Scrap, unlike the other raw-material inputs, is not restricted in its location. It is residual both of industrial activities and of the daily activities of the population.[7]

In estimating the availability of scrap, one must take into account three basic sources of supply: internal scrap, which is the outcome of the very process of making steel and of rolling operations; manufacturing scrap, generated in products made from steel (this residual can reach up to 35 percent of the total steel material used) ; obsolescent scrap, derived from obsolescent ships, cars, equipment, and other material made from steel.

Steel experts in Brazil have devised rule-of-thumb methods to estimate the availability of scrap formed by the above-mentioned processes.[8] Es-

7. Our discussion of the generation and consumption of scrap in Brazil is based on the work of J. M. Falcão in the CEPAL/Falcão Study *A Economia Siderurgica da America Latina: Monografia do Brasil,* pp. V 33–43; and the work of the Brazilian consulting firm Tecnometal done for the BNDE entitled "Mercado de Sucata de Aço," internal report of BNDE, 1965.

8. See the final section of this chapter for detailed estimates.

timates based on these methods have shown that internal scrap produced ranges between 20 to 25 percent of the weight of the ingots used and that manufacturing scrap has been estimated to be generated at a rate of 8 to 13 percent of the weight of rolled steel products consumed. The estimation of generating obsolescent scrap must be even more arbitrary. One estimate used is based on the hypothesis that there is an average lifetime for steel consumed (in any form) in a given year. The calculations reproduced in the final section of this chapter assume a lifetime of fifty years; it was thus estimated that the yearly scrap formation was equal to one fiftieth of the steel consumed up to the year in question.

The data reproduced in the final section of this chapter show that up to 1957 the yearly generation of scrap was greater than the yearly consumption by the steel industry. Since 1958, however, yearly scrap consumption has been greater than yearly generating. In 1964, for example, 1.14 million tons of scrap were generated, while consumption amounted to 1.45 million tons. Projections show that by 1970 scrap generated will be about 1.95 million tons and consumption 2.52 million tons.

Until 1957, Brazil's scrap reserves had continuously accumulated, reaching an estimated 5.6 million tons. Since then the excess of consumption has reduced reserves to 4.2 million tons in 1965, and one projection shows that accumulated scrap reserves will be reduced to 1.7 million tons in 1970.[9]

The conclusion reached from these projections is that in the 1970s Brazil will have to import scrap. It is possible, however, that these projections are a little too pessimistic. As will be seen in a later chapter, a large proportion of Brazil's steel expansion program is based on LD furnaces with very low consumption of scrap. Also, the modernization of existing SM furnaces will result in a proportionally lower input of scrap. These developments would, of course, reduce the rate of growth of scrap consumption.[10]

ELECTRIC POWER

Both actual and potential hydroelectric power resources in Brazil are favorably located in relation to the industrial heartland of central-south Brazil and to the major steel-producing centers. The great escarpment

9. This is based on the Tecnometal projections. See footnote 7.
10. Substantially more scrap could be obtained in Brazil by modernizing the organization of scrap collection.

that stretches several hundred miles north and south of Rio de Janeiro on the Atlantic coast, containing many rivers, offers first-rate sources of hydroelectric power. The various hydroelectric works erected in this area, combined with a huge hydroelectric system being erected on the Paraná river, should provide adequate power reserves for the 1970s and 1980s.

Table 12

Actual and Planned Power Capacity of Brazil

(capacity in megowatts)

	1965	1966	1967	1968	1969	1970/6	Total	Grand Total by 1976	1965 watts p/cap
North	87	53			60		113	200	29
Northeast	721	4	248	46	36	75	403	1,130	29
Central/south	5663	90	567	510	1310	2540	5017	16,680	158
South	857	120	150	18	45	80	413	1,270	58
Central/west	82	15	13	60	50		138	220	20
Total	7410	282	978	634	1501	2695	6090	13,500	90

Source: APEC, Análise e Perspectiva Econômica, *A Economia Brasileira e Suas Perspectivas.* Rio de Janeiro: APEC Editora S.A., 1967.

All regions, except the Amazon area (which is the "north" area in Table 12) and the south, have adequate hydroelectric reserves. The expansion program of the 1960s should double the installed capacity of 1965 by the mid-1970s.

It has been found that despite the favorable power picture rates in the mid-sixties were too high. In 1966 power costs in selected cities were (cost of 1,000KWH)

Belo Horizonte	US$ 10
Rio de Janeiro	18 to 30
São Paulo	15 to 20
Vitória	24
Pôrto Alegre	14
Recife	19

When these rates were obtained there was an imminent increase in Belo Horizonte of US$ 5 and a decrease in Vitória of US$ 7. A 5 percent tax to cover the costs of new facilities contributes to the high costs of power.

The above survey has shown that Brazil has ample natural resources for the production of steel. The exception is the availability of good coking coal. Given the country's immense iron ore reserves, however, it would seem economical for the country to continue to promote the export of iron ore in exchange for good coking coal. As a matter of fact, it is obvious that the size and quality of Brazil's iron ore reserves are so vast that they can serve both as an earner of foreign exchange for the importation of coal and as an earner of foreign exchange above what is needed to finance the necessary coal imports.

I have not discussed the country's transportation system in order to see how efficiently these materials can be assembled for the production of steel. I defer this discussion until a later chapter which will examine the locational pattern of the Brazilian steel industry. Also, a discussion of the availability of human resources for the production of steel will be included in a general consideration of the industry's historical development.

SUPPLY AND DEMAND OF SCRAP IN BRAZIL

The two tables that follow contain two independent estimates of the demand for and the supply of scrap in Brazil.

Table 13 contains the estimates made by J. M. Falcão for the Economic Commission for Latin America.[11] The estimate of internal scrap is based on the assumption that on the average 22.5 percent of the weight of ingot production is recoverable. It was not possible to obtain estimates of steel production in a systematic way before 1925, hence the estimates of the generation of internal scrap start only in that year.

It is even more difficult to estimate the availability of manufacturing scrap. It is known that some manufacturing processes generate as much as 35 per cent of the total weight of the material as scrap. The dearth of data on the consumption of steel in various manufacturing processes in Brazil makes such estimates very difficult. United States estimates show that the rate of scrap recovery in manufacturing varies between 8 and 13 percent. Falcão employed the more conservative estimate of 8 percent of the weight of total steel (rolled steel) consumed in manufacturing.

In order to estimate obsolescence scrap, it was assumed that an average period of 50 years would elapse for a newly produced steel product to go to the scrap pile. Since scrap formation is a continuous process it was

11. J. M. Falcão's estimates can be found in the mimeographed work he did for the Comissão Econômica Para A América Latina, *A Economia Siderúrgica da América Latina: Monografia do Brasil*, pp. V 33–43.

assumed that usable scrap each year corresponded to one fiftieth of total steel consumed until that year. Data on consumption of rolled steel products were available only from 1901 on.

Table 13 summarizes the data. Until 1925, accumulation is based only on obsolescent scrap. The estimates had to overcome two obstacles. First, the generation of scrap takes place over a vast area, while its consumption is concentrated in just a handful of states (São Paulo, Rio de Janeiro, Minas Gerais, Rio Grande do Sul). Taking into account the transport difficulties and a rather backward system of scrap-collection, one must assume some losses in actual scrap availability. Second, some account must be taken for the scrap directly consumed by small and widely dispersed foundries. To overcome these difficulties, Falcão has assumed that 15 percent of the scrap generated is not available to the steel industry. Hence an adjustment has been made, as seen in Table 13 (see columns D, E, and F).

Scrap demand estimates were arrived at in the following manner. For the period 1925–40 it was estimated that the consumption of scrap and pig iron by steel shops was equal; that is, the over-all input proportions between the two was about fifty-fifty. In the forties, with increased production of pig iron, the proportion of scrap used fell to 40 percent (these calculations are made from 1941 on). After 1965, with the increased use of LD shops where scrap consumption is low, it was assumed that scrap would amount to only 35 percent. Applying these proportions to pig iron–consumption data, the demand for scrap was obtained. Assuming a further diminution of the scrap content to 30 percent in 1975, demand and supply should be about equal at that time.

Table 14 contains an independent estimate of scrap generating and demand made by the Brazilian consulting firm Tecnometal. It was estimated that apparent scrap return of domestically produced and imported steel is about 70 percent of apparent consumption over a 28-year cycle. Thus, over the ten years since 1954, about 290,000 tons were available.

Table 13

Brazil: Total Scrap Generation for Steel Industry

(in thousands of tons)

Year	Internal scrap A	Manu-facturing scrap B	Obso-lescent scrap C	Total[a] scrap D	Unused[b] scrap E	Scrap available F	Total scrap consumption by steel industry G	Annual net for accumula-tion H	Total accumulated scrap reserves I
1924									797[c]
1925	2	30	81	113	17	96	5	91	888
1926	2	32	86	120	18	102	6	96	984
1927	2	35	91	128	19	109	5	104	1 088
1928	5	39	97	141	20	121	12	109	1 197
1929	6	41	103	150	22	128	15	113	1 310
1930	5	21	106	132	19	113	12	101	1 411
1931	5	12	108	125	18	107	13	94	1 505
1932	8	13	110	131	18	113	19	94	1 599
1933	12	22	113	147	20	127	30	97	1 696
1934	14	28	117	159	22	137	34	103	1 799
1935	14	28	120	162	22	140	35	105	1 904
1936	17	31	124	172	23	149	41	108	2 012
1937	17	40	129	186	25	161	42	119	2 131
1938	21	28	133	182	24	158	51	107	2 238
1939	26	34	137	197	26	171	63	108	2 346
1940	32	33	141	206	26	180	78	102	2 448
1941	35	29	145	209	26	183	68	115	2 563
1942	36	21	148	205	25	180	70	110	2 673
1943	42	25	151	219	27	192	82	110	2 783
1944	50	39	155	244	29	215	97	118	2 901
1945	46	37	158	241	29	212	91	121	3 022
1946	77	35	164	294	33	261	151	110	3 132
1947	87	59	170	316	34	282	170	112	3 244
1948	109	45	174	328	33	295	212	83	3 327
1949	138	56	180	374	35	339	271	68	3 395
1950	178	67	187	432	38	394	347	47	3 442

The Brazilian Steel Industry

Table 13 (continued)

Year	(A)	(B)	(C)	(D)	(E)	(F)	(G)	(H)	(I)
1951	190	85	195	470	42	428	371	57	3 499
1952	201	87	204	492	44	448	393	55	3 554
1953	229	81	212	522	44	478	447	31	3 585
1954	258	119	222	599	51	548	505	43	3 628
1955	270	101	231	602	50	552	528	24	3 652
1956	320	106	240	666	52	614	626	− 12	3 640
1957	347	122	251	720	56	664	678	− 14	3 626
1958	394	121	262	777	57	720	770	− 50	3 576
1959	442	160	276	878	65	813	864	− 51	3 525
1960	499	170	291	960	69	891	976	− 85	3 440
1961	547	181	306	1 034	73	961	1 070	−109	3 331
1962	569	182	322	1 073	76	997	1 113	−116	3 215
1963	633	218	341	1 192	84	1 108	1 237	−129	3 086
1964	731	240	362	1 333	90	1 243	1 430	−187	2 899
1965	900	263	385	1 548	97	1 451	1 540d	− 89	2 810
1966	1 015	289	411	1 715	105	1 610	1 736	−126	2 684
1967	1 078	314	438	1 830	113	1 717	1 844	−127	2 557
1968	1 238	344	468	2 050	122	1 928	2 117	−189	2 368
1969	1 318	375	501	2 194	131	2 063	2 256	−193	2 175
1970	1 341	410	537	2 288	142	2 146	2 295	−149	2 026
1975	2 225	638	593	3 456	185	3 271	3 264	7	

Source: Comissão Economica Para a America Latina, "A Economia Siderurgica Da America Latina: Monografia Do Brasil," Prepared by J. M. Falcão, p. V-42.

a Total Scrap: (A) + (B)% + (C)
b Unused Scrap: 15% of (B) + (C)
c Reserves accumulated since 1901, having deducted 15% not usable in steel industry.
d Production of 1965–70 based on projections by Falcão.

Table 14

Brazil: Tecnometal/BNDE Estimates of Scrap Generation and Demand

(in thousands of tons)

Year	Internal scrap	Manufacturing scrap	Obsolescent scrap	Total scrap	Total consumption	Net accumulation	Estimated reserves
1952							5 100
1953	210	75	290	575	380	195	5 300
1954	240	105	290	635	465	170	5 500
1955	250	90	290	630	555	75	5 500
1956	250	95	290	665	655	10	5 600
1957	305	105	290	700	700	0	5 600
1958	345	105	290	740	785	−45	5 500
1959	385	180	290	855	880	−25	5 500
1960	465	190	290	945	1 060	−115	5 400
1961	470	200	290	960	1 065	−105	5 300
1962	530	210	290	1 030	1 210	−180	5 100
1963	580	235	290	1 105	1 335	−200	4 900
1964	630	225	290	1 145	1 450	−305	4 600
1965	770	270	290	1 330	1 680	−350	4 200
1966	890	300	290	1 480	1 890	−410	3 800
1967	990	320	290	1 600	2 300	−490	3 300
1968	1 100	350	290	1 740	2 300	−560	2 800
1969	1 140	380	290	1 810	2 310	−500	2 300
1970	1 240	420	290	1 950	2 520	−570	1 700

Source: Unpublished report of Tecnometal for the Banco Nacional de Desenvolvimento Economico (BNDE).

4

HISTORICAL DEVELOPMENTS: COLONIAL TIMES TO THE PRESENT

GIVEN THE vast natural resources for making pig iron and steel examined in the previous chapter, it is not surprising to discover that attempts to produce iron and iron products go back to colonial time. There are two reasons for making an extensive historical survey of iron and steel production in Brazil. First, it should help to make it clear that the idea of iron and steel making in Brazil has long historical antecedents and is not the brainchild of latter-day, development-conscious planners. Second, it helps to show the various institutional barriers that prevented the earlier appearance of iron and steel making on a large scale.

DEVELOPMENTS BEFORE THE 20TH CENTURY

The earliest evidence of iron-working activities appears in a report by Father José Anchieta to his Jesuit superiors and to the king of Portugal, mentioning the existence of iron ore deposits in the interior of the captaincy of São Vicente (now the state of São Paulo).[1] In 1556, the Jesuit Mateus Nogueira established a forge to produce fishhooks, knives, wedges, shovels, and other implements for his community. These are believed to be the first Brazilian-made iron implements.[2] By 1586, numerous forges and blacksmiths existed in the central interior region of Brazil.

1. Much of the material on the early Brazilian iron industry is based on Edward J. Rogers, "The Iron and Steel Industry in Colonial and Imperial Brazil," *The Americas*, October 1962; See also Heitor Ferreira Lima, "Indústrias Novas No Brasil—Siderurgia No Passado." *Observador Econômico e Financeiro*, ano XXIII, n. 264 and 265; and Geraldo Magella Pires de Mello, "Histórico, Possibilidades e Problemas Da Siderurgia no Brasil," *Observador Econômico e Financeiro*, No. 262.
2. Edward J. Rogers, *op. cit.*, p. 172.

It is generally agreed, however, that the iron and steel industry was initiated in 1597 at Biraçoiaba in the captaincy of São Vicente (near the present town of Sorocaba in the state of São Paulo) by Affonso Sardinha Filho. The iron ore deposits were actually discovered by Sardinha's father, but it was he who constructed two Catalan forges that turned out the first commercial iron ever produced in Brazil.[3] These forges continued to produce until Sardinha's death in 1629. Sardinha was also involved in the construction of a number of other small smelters which were encouraged and partially subsidized by Francisco de Souza, the governor of the captaincy of São Vicente. In fact, this was the first attempt in Brazil to develop an iron-working industry.

The early interest in iron-smelting faded after 1629, and until the next century little was done to promote further this industry. Here and there a few forges were founded, but most of them had a very short period of operation.[4] Of course, iron-smelting never stopped. In many parts of the central-south of Brazil, products were being smelted by blacksmiths using very primitive methods. Many of the techniques used by the blacksmiths were learned from African slaves.[5]

The discovery of gold in the present state of Minas Gerais in the middle of the eighteenth century led a number of Brazilian administrators to petition Portugal for permission to build smelting furnaces in order to make iron implements necessary in the mines. The argument in favor of such action was that it was cheaper to produce these implements in Brazil than to import them, lowering the cost of mining and thus increasing profits. The reaction of Portugal was negative, since this would be con-

3. Rogers mentions, "Anvils made here can still be seen in Brazil and metallurgists have attested to the quality of the workmanship." *Op. cit.*, p. 173.

4. It has been claimed by contemporary experts that the reason for the early failures of these operations was the presence of titanium in the São Paulo ores of Ipanema, close to Sardinha's furnaces. Titanium is difficult to separate from iron ore because of titanium's high melting point.

5. Rogers mentions that the "contribution of the African slave to Brazilian metallurgy should not be overlooked. There is a school of thought which implies that the art of metallurgy in the Western world may have begun in Middle Africa. It is asserted that these Middle African techniques may have been absorbed by the Moors of North Africa and then passed on to Europe by them during their conquests there. Suffice it to say here that it was the African slave who introduced into São Paulo and Minas Gerais the iron working process known as *de cadinho* (literally means 'by crucible' or melting pot) which can still be seen in operation in remote areas of Brazil." Rogers, *op. cit.*, p. 174. Rogers also gives as his principal source for these facts: Forbes, Robert H., "The Black Man's Industries," *Geographical Review* XXIII, No. 2 (April 1933) 230–236.

trary to its mercantilistic policies. It was feared by Portuguese authorities that permission for such developments would be the first step towards an economic independence of the colony. Also, it was feared that such industries would draw workers from other government enterprises.

The attitude of Portugal did not change until the very end of the eighteenth century. For example, in 1780 a request of the governor of the captaincy of Minas Gerais, Rodrigo José de Menezes, to permit the establishment of smelting plants was turned down, and early in 1785 the Portuguese government signed an order making the existence of such operations illegal and ordering the destruction of all existing furnaces. The same order requested the Brazilian administrators to "attend to agricultural interests and the mining of gold."[6] Portugal's control, however, was not very thorough, and in Minas Gerais, São Paulo, and other places the smelting of iron *em cadinho* continued.

With the accession of Prince João, later King João VI, as regent of Portugal, policies became more liberal. This was reflected on May 27, 1795, when the governor general of Brazil, Louis Pinto de Souza, informed the governors of the captaincies that iron-smelting and the establishment of foundries was once again permitted. This resulted in the construction of a number of small furnaces and foundries throughout the countryside. In 1800, Colonel Candido Xavier de Almeida was asked by colonial authorities to travel with two experts to Sorocaba to study the feasibility of building an iron smelter in nearby Ipanema. The report of the group was favorable, and in April 1801 the founding of an iron industry in the Sorocaba region was ordered. Construction began almost immediately, but many years passed before iron was actually smelted.

The arrival in 1808 of the royal family in Brazil gave a new impetus to various industries, including iron-smelting. Prince João ordered the acceleration of construction on the Ipanema smelter and advanced government funds to construct a blast furnace and refining forges at Morro do Pilar (also known as the Real Fábrica do Morro Gaspar Soares) in Minas Gerais. The construction of the latter was to be supervised by the Intendente Manoel Ferreira da Camara Bittencourt e Sá.[7] Also, in May 1810 the Government lent funds for the construction of a foundry specializing in artillery pieces, and in December of that year a company was established by royal order to mine iron ore in the captaincy of São Paulo.

6. Rogers, *op. cit.*, p. 175.

7. Usiminas named its main plant after this Brazilian iron-making pioneer "Usina Intendente Camara."

Though directed by the able Intendente Camara, the works in Minas Gerais at Morro do Pilar proceeded with difficulty. In July 1814, one of the walls of the blast furnace collapsed. Finally, in 1815, the works were completed. It included the blast furnace, twenty-eight feet high and with a maximum width of three feet, several refining furnaces and three Catalan forges to be used as auxiliary units in case of a failure by the blast furnace. The blast furnace was the first to function in Brazil. Much of the production of the Morro do Pilar was destined for the diamond industry of Minas Gerais. One source offers the following data on the annual output of iron for the first seven years of the plant's production:

1815	395 *arrôbas*[8]
1816	1,156
1817	796
1818	936
1819	701
1820	2,536
1821	346

Another source claims that the total output from 1815 to 1831, from the opening of the ironworks until its closing, attained 8,905 *arrôbas*, about 133 tons.[9] During its lifetime, the Morro do Pilar works encountered many difficulties. Shortages of water often paralyzed the operations for two or three days a week. It was discovered that a father-and-son team of artisans imported from Prussia were ignorant of the functioning of such operations. Wages of workers were often unpaid for long periods of time and consequently many left. The works operated fitfully throughout the 1820s, and in 1831 they closed, the equipment being sold at public auction. Thus ended Brazil's first attempt to run a relatively large, for its time, ironworks—also Brazil's first successful blast furnace.

The construction of the Ipanema works continued to drag on until 1814, when the authorities appointed a new director, the metallurgist Frederico Guilherme de Varnhagen.[10] It took him until 1817 to finish the

8. Lima, *op. cit.*, No. 264. The arrôba is an old Portuguese weight which was used in Brazil. It is equal to about 32 pounds.

9. Rogers, *op. cit.*, p. 179

10. Varnhagen served Portugal as a metallurgist from 1803 on; he came to Brazil in 1808 with the Portuguese court as a first sergeant in the Royal Corps of Engineers. He was born in Germany and returned there after long years of service to Portugal and Brazil.

works; many quarrels occurred over methods of operation. The works were further delayed by difficulties in the entry of European artisans engaged for the operation of the plant. In November 1818, three years after the opening of Morro do Pilar, with the help of only local skilled labor, the first iron was poured at Ipanema.

The Ipanema works had a long and checkered history. It was managed by Varnhagen until his return to Europe in 1821.[11] After 1824, its operations steadily declined in efficiency, and in 1832 the works were shut down. It should be mentioned at this point that both Morro do Pilar and Ipanema were relatively high-cost operations and had difficulties in competing against British-made iron products, which had free access to the Brazilian market under treaty rights granted to England by João VI.

The Ipanema works were reopened a number of times during the nineteenth century. They were operated from 1836 to 1842 under the direction of a Major João Bloen. The latter was arrested in 1842 for revolutionary activities, and the works were closed again. They were reopened intermittently and operated by army officers, but in 1864 they were ordered closely permanently by the government because the equipment was considered obsolete. The necessities of the Paraguayan War, however, once again forced the opening of the Ipanema plant, which was renovated for the occasion. Operations continued until the 1890s, efficiency declining drastically again in the two final decades. In 1895, the Brazilian Congress ordered the plant closed for the last time "thereby bringing to a close an operation which had almost become an institution in Brazil."[12] A long time was to pass until São Paulo would again have a large iron works.

This historical survey should also mention the activities and contributions of the Baron Guilherme de Eschwege, who came to Brazil with the royal family as a specialist in the field of metallurgy. Eschwege was born in Germany in 1777 and remained in Brazil until 1821, when he returned to Portugal with the royal family. In Portugal, Eschwege had already won the admiration of the Portuguese authorities by directing iron-smelting and fabricating in that country. In Brazil he was named "Director of the Gold Mines and Trustee of the Cabinet of Mineralogy of the

11. It has been estimated that between 1818 and 1821 the total production of the Ipanema works amounted to
> 16,085 *arrôbas* of iron bars
> 18,087 *arrôbas* of pig iron
> 12,589 *arrôbas* of "obras fundidas"

Lima *op. cit.*, No. 264.

12. Rogers, *op. cit.*, p. 181.

Government" and was supposed to be in charge of "examining the mineralogical products of the captaincy, opening mines, constructing metallurgical plants, principally those of iron. . . ."[13]

In addition to his vast contributions to Brazilian geological studies,[14] Eschwege led the construction of another pioneer iron works at Congonhas do Campo, Minas Gerais. The equipment used was older than that used in the Morro do Pilar or the Ipanema plant (Catalan forges instead of a blast furnace were used); however, Eschwege made some innovations, especially by using hydraulic power for operating the equipment. The works were built along the Rio da Prata, close to a waterfall where a farmer had already built a waterwheel. The waterwheel powered the mechanical hammers used to crush the ore and shape the iron bars and operated the bellows in the ore-smelting process. The plant, named Patriótica, produced its first iron bar on December 12, 1812.

The Eschewege operations prospered and were even expanded. The production record of the plant was estimated at

1813	996 *arrôbas*
1814	997
1815	1,278
1816	1,134
1819	1,643
1820	1,229[15]

The old Catalan forge technique of melting iron was not, however, competitive with imported British goods produced by newer methods. In the 1820s, the firm lost more and more of its market, and shortly afterward the whole enterprise failed. The ruins of the works may still be seen in Minas Gerais.

In 1817, a young French engineer, Jean Antoine de Monlevade, arrived in Brazil. He was a graduate of the Polytechnic School of Paris, finishing his studies in 1815 as the second highest student in his class. In compensation for his scholastic achievements, the French government

13. Rogers, *op. cit.*, p. 177.
14. Macedo Soares has even called him "o pai de nossa Geologia" (the father of our geology), see Edmundo De Macedo Soares e Silva, "Desenvolvimento Da Siderurgia No Brasil Nos Ultimos Vinte Anos," *Metalurgia—ABM*, XXI, No. 86, (1965), 5.
15. Lima, *op. cit.*, No. 264.

awarded Monlevade a travel grant for the purposes of extending his studies. Monlevade chose to go to Brazil.[16]

At the beginning, Monlevade worked with Eschwege in a number of enterprises, including the installation of the Catalan forges at Congonhas. After some time, he left Eschwege to strike out on his own and with a Captain Luis Soares de Gouveia he built a blast furnace near the village of Caeté in the interior of Minas Gerais. His site, not far from the present city of Belo Horizonte, is the location of the present firm Companhia Ferro Brasileiro. Monlevade's furnace went into production in 1818. For some obscure reason, the operation was soon abandoned.

Monlevade ventured further into the interior of Minas Gerais. At São Miguel de Piracicaba, he selected a site for his new iron works. This site was an estate owned by the Province of Minas Gerais, which he in time acquired from the provincial government. Instead of a blast furnace, he built some Catalan forges and taught his slaves the art of iron-making.[17] The first iron was produced there in 1825. The operation gradually became the most important in the region, producing tools and implements for agriculture and for the gold and diamond-mining activities. The operation prospered during Monlevade's lifetime, but after his death production declined and the enterprise was abandoned.

The death of Monlevade and the return of Eschwege and Varnhagen to Europe ended these early attempts at a systematic establishment of an iron-making industry. The reason for the decline of these establishments was caused, not only by the disappearance of leadership, but also by British competition from which the government could not protect domestic enterprises.[18] In addition, labor was scarce, being siphoned off by the mining of gold, diamonds, the production of rubber, and later coffee. It has been claimed that the dearth of qualified labor forced some enter-

16. Rogers tells an interesting anecdote about the reason for Monlevade's ultimate permanence in Brazil. "The voyage from France was made . . . by sailing vessel. The crossing was unusually rough and Monlevade, fortunately for Brazil, proved to be a poor sailor. It is said that this disagreeable experience caused him to lose any desire for a return crossing. . . ." *Op. cit.*, p. 182.

17. The use of slaves in such operations was common in those days. We already mentioned the iron-making techniques imported from Africa. It was reported by one observer that the Ipanema works in its periods of operation employed 85 slaves and 24 free men. See Lima *op. cit.*, No. 264.

18. In exchange for the help João VI received from the British (the royal family sailed to Brazil in its flight from Napoleon under British protection), Britain received special commercial privileges in Brazil.

prises to revert to more primitive methods of production, thus putting them in an even worse competitive position.

Iron-smelting and fabricating did not disappear from Brazil as the nineteenth century wore on. According to a report by Eschwege, there existed in 1821 about thirty forges in Minas Gerais, with a daily output of 100 to 400 arrôbas each—about 120 tons a year. In 1864, 120 forges were in operation, producing 1,550 tons annually. The methods used, however, were primitive. Of thirty iron works reported in the headwater region of the Rio Doce in 1879, seven used Italian forge methods and the rest used the old African *cadinho* technique.

An important event in the latter part of the nineteenth century for future developments in the steel industry was the foundation in 1879 of the School of Mines at Ouro Preto, Minas Gerais, by the French engineer Henrique Gorceix. The latter had been chosen personally by the Emperor Dom Pedro II to organize a school for training geologists and metallurgical engineers. Almost from the very beginning the school furnished some of the country's best engineers. A little later, the founding of the Escola Politécnica in São Paulo was also of great importance. The school gave engineers a rigorous training and some the opportunity of specializing in metallurgy. The school had laboratories and a foundry for the study of metallurgy.

It has been noted in passing that the military always had an interest in the development of iron-working and later of a steel industry. We have seen that during its existence the Ipanema plant produced largely military equipment. The interest of the military in the development of iron and steel production continued throughout the nineteenth and twentieth centuries. The reason for this interest is obvious, given military needs. The continued interest of the military in industrial development in general and in the steel industry in particular resulted in the founding of the Escola de Engenharia do Exército (also called Escola Técnica do Exército) in 1930. The school's objective was to train engineers, including metallurgical engineers. Many of the graduates of this school were among the leaders of the steel industry as it developed from the 1930s to the 1960s.

DEVELOPMENTS IN THE FIRST THREE DECADES OF THE TWENTIETH CENTURY

Iron production grew in the last years of the nineteenth century and in the early part of the twentieth century at a steady but slow pace. Most of the output was produced in small shops and foundries which made spare parts for the railroads, the military, or to service the machines and provide utensils for the large coffee plantations and sugar mills.

The renewed interest in iron and steel production created by the school at Ouro Preto stimulated a substantial amount of research into new techniques of production, and in 1888 the first new blast furnace since the failures of the early part of the century was established. The plant, called Esperança, was built by three entrepreneurs (Joseph Gerspacher, Amaro da Silveira, and Carlos da Costa Wigg). It was located near the town of Itabirito, Minas Gerais. The blast furnace had a daily capacity of six tons of pig iron which was used for foundry work. The furnace used charcoal from the company's own lands, and iron ore came from the deposits of the Itabira Mountain, which the company also owned. In 1893, the founders of Esperança built another small plant in the town of Miguel Burnier in Minas Gerais.[19]

In 1899 these works were acquired by the engineer J. J. Queiroz, and under this management they entered a long period of prosperity. In 1908, Queiroz added a second furnace to the Esperança plant. By 1915, the latter had a daily capacity of twenty-five tons and the Burnier plant fifteen tons, and the firm employed 412 laborers.[20]

According to one authority, Brazil entered the twentieth century producing about two thousand tons of pig iron in about seventy small establishments.[21] By the end of the first decade of the twentieth century pig iron production had not grown much beyond that amount, while imports

19. Gonsalves, Alpheu Diniz, *et al.*, *O Ferro na Economia Nacional*, pp. 22–23; Elysio de Carvalho, *Brasil: Potência, Mundial*, p. 170.

20. After the death of Queiroz, the works were named after him. By 1915 the pig iron produced was used partly within the plant for the production of pipes for irrigation, parts for sugar mills, and agricultural implements. It seems that the quality of the pig iron was very good and the works won gold medals in the St. Louis and Buffalo international expositions. Production statistics of the firm are as follows (in tons):

1899	79.6	1904	1,710.0	1909	2,133.9	1914	2,181.0
1900	756.0	1905	1,394.0	1910	2,658.5	1915	3,279.3
1901	825.5	1906	1,654.0	1911	3,261.6		
1902	1,258.2	1907	1,900.6	1912	3,463.1		
1903	1,359.5	1908	1,868.3	1913	3,999.7		

Carvalho, *op. cit.*, p. 173.

As mentioned in the text, the Esperança works were first set up by Gerspacher, Silveira, and Wigg. About two years after the founding of the works, the Companhia Forjas e Estaleiros acquired possession. In 1899, Queiroz bought the installations from the CFE group. See Gonsalves, *op. cit.*, p. 22.

21. Humberto Bastos, *A Conquista Siderúrgica No Brasil*, p. 70.

of rolled steel products averaged 272,500 tons yearly in the periods 1908–12. In 1916, pig iron production amounted to only 4,267 tons, and in 1924 (the earliest date from which we have a systematic accounting of steel production) total output of steel amounted to only 4,492 tons.

Brazil's growing steel consumption could at the time be satisfied only by increasing imports. Government circles thus began to realize that the alternative, a substantial expansion of domestic steel production, could only be attained with the co-operation of foreign capital and/or special inducements for pooling domestic savings for the creation of a large domestic industry. In 1909, the government of President Nilo Peçanha offered special monopoly privileges and subsidies to induce either foreign or domestic capital to establish a large-scale steel industry.[22] In 1911 the engineer Trajano de Medeiros presented a project for the establishment of an integrated steel mill (with 150,000 tons capacity) which was to be located in Juiz de Fora, Minas Gerais. The plans came to nothing, however, because the Balkan War and World War I closed the international capital markets.[23]

Although the First World War acted as a stimulant to Brazilian industrial growth, resulting in a substantial expansion of many light industries, especially textiles, its effect on the iron and steel industry was not very pronounced. Some increase in output in existing blast furnaces in Minas Gerais occurred. Pig iron production, which was established at about three thousand tons in 1914, rose to more than ten thousand tons in 1919. As mentioned above, however, steel production in the immediate post–World War I period was still negligible. The output during this period was destined largely for small foundry production. By the mid–1920s, almost all of Brazil's consumption of rolled steel was still imported.[24]

Although the decade of the twenties was characterized by a decline in

22. Among the types of privileges offered by the government were government guarantee of annual steel consumption amounting to ⅓ of each firm's capacity, tax exemptions, guaranteed lower transportation rates on railroads. See Carvalho, *op. cit.*, p. 146; also Ministério da Viação (Brasil), *Revisão do Contrato da Itabira Iron*, p. 39.

23. The principal source for the following narrative was: John D. Wirth, "Brazilian Economic Nationalism: Trade and Steel Under Vargas," unpublished doctoral dissertation, Stanford University, March 1966; see also Edmundo de Macedo Soares e Silva, "Desenvolvimento Da Siderurgia No Brasil. . . ." *Op. cit.;* Bastos, *op. cit.;* Geraldo Magella Pires de Mello, "Histórico, Possibilidades e Problemas da Siderurgia No Brasil," *Observador Econômico e Financeiro,* No. 262, pp. 56–64.

24. See Roberto C. Simonsen, *Brazil's Industrial Evolution,* pp. 54–5; also Werner Baer, *Industrialization and Economic Development in Brazil,* pp. 15–20.

the rate of growth of industrial activity (through renewed European and American competition much idle capacity developed in the Brazilian textile industry),[25] the iron and steel industry continued to grow, and by the second half of the twenties steel-making began to expand. Steel ingots, which before World War I were still largely imported, were almost fully supplied by Brazilian firms in the early twenties (though production was still relatively small). In the second half of the twenties there was also some small beginning in building rolling-mill capacity, and by the beginning of the thirties Brazilian firms supplied more than 10 percent of total consumption of rolled steel products (See Table 15 for details).

The growth of iron and steel production in the first three decades of the twentieth century was based entirely on the initiative of private entrepreneurs. The main production center until the end of the second decade was in Minas Gerais, where the dominant producer was the firm Queiroz Junior. This firm was responsible for most of the pig iron production increase in the World War I period.

During the war, two engineers, Amaro Lanari and Cristiano Guimarães, founded a firm which began producing in 1917—the Companhia Siderúrgica Mineira. This company seems to have experienced many early difficulties. By a fortunate coincidence, in 1920 when the company was close to collapse King Albert of Belgium made an official visit to Brazil. While visiting the state of Minas Gerais, Governor Artur Bernardes tried to interest the king in the possibilities of investing Belgian capital in the state, especially in steel production. The Belgian government and private Belgian groups seem to have been interested, and in 1921 a group of Belgian representatives headed by the engineer Jean Pierre Arend of the ARBED (Acieries Réunies de Burbach-Eich-Dudelange) steel group visited Minas Gerais. Guimarães, the co-owner of the Cia. Siderúrgica Mineira (CSM), who also happened to be in charge of the Belgian consulate in Belo Horizonte, suggested at that time that instead of founding a new firm the ARBED group should associate itself with or even absorb the already-functioning CSM. The suggestion was accepted, and in December 1921 the Cia. Siderúrgica Mineira became the Companhia Siderúrgica Belgo-Mineira.[26]

25. In the twenties the Brazilian authorities, concerned mainly about Brazil's traditional exports, failed to take any actions to protect the country's infant industries against the onslaught of better known and better quality foreign products. See Baer, *op. cit.*, p. 19.

26. Bastos, *op. cit.*, p. 111.

From the beginning, Belgo-Mineira[27] committed itself to building an integrated charcoal-based steelworks at Monlevade, the site where the engineer Monlevade had built his enterprise in the nineteenth century. Construction of this plant could not, however, be undertaken until a rail link from Belo Horizonte was completed. This did not occur until 1934. The company meanwhile operated and expanded the works that the old firm had already established at Sabara, close to Belo Horizonte. The Sabara works, which initially only consisted of a blast furnace, were gradually enlarged into a small but integrated plant with the addition of a small SM steel furnace and a small rolling mill.

It is interesting to note that most of the small iron and steel plants that appeared in the twenties were constructed by firms whose principal activity was in other fields. The steel plants usually complemented these other activities. In 1919, Aços Paulista was created in São Paulo to produce machinery for mining; a steel foundry was built in 1923 in order to produce components and spare parts for these machines. In 1920, the Dedini group, which produced machines and assorted equipment for the growing sugar industry, founded a steel firm, M. Dedini S.A. This new firm consisted of a steel and iron foundry, producing some of the principal pieces for the parent company. It built an electric furnace and an SM furnace. Some of the other small firms founded in this period produced rolled nonflat products with fairly primitive rolling mills, especially bars and rods for the construction industry, and assorted shapes.[28]

It should be noted here that although the government had a generally passive attitude towards industry in the interwar years, this was not the case in the iron and steel industry. Government decrees in 1918 and again in 1925 gave firms producing iron and steel various types of tax favors, better credit conditions, lower freight rates, and import duty exemptions.[29] In 1922, many types of special favors were granted the Usina Queiroz Junior.[30]

Brazilian output of pig iron by 1930 was 35,305 tons, while pig iron imports amounted to almost 2,000 tons, having been substantially higher

27. To be exact, the ARBED group consisted of Belgium and Luxemburg capital.
28. Some of the other firms founded in that period and still producing today are Indústria Metalúrgica Souza-Noschese in São Paulo (1920); Siderurgica Aliperti, São Paulo (1924); Companhia Brasileira de Usinas Metalúrgicas (1925); Usina Santa Olímpia Limitada (1925).
29. Gonsalves, *op. cit.*, pp. 113, and 123.
30. *Ibid.*, p. 115.

throughout the decade of the twenties. One leading authority on the Brazilian steel industry, General Macedo Soares, has estimated that pig iron–making capacity in 1930 was equal to 100,000 tons a year.[31] A substantial amount of the pig iron produced was not sold or consumed directly because of its poor quality, and throughout the twenties foundries had a preference for imported pig iron. It was also difficult in the twenties to sell local foundry products. Not only was local pig iron or steel output of poor quality, but such techniques as using sand and making molds were still not mastered.

It will be seen in Table 15 that the decade of the thirties was much more dynamic for the iron and steel industry. Pig iron production increased more than fivefold to more than 185,000 tons in 1940, while imports of pig iron almost completely disappeared. Ingot steel production increased about sixfold, from 20,985 tons in 1930 to 141,201 tons in 1940, imports almost disappearing.[32] A substantial rate of increase in rolled-steel production can also be observed. With the industrialization spurt of the 1930s, however, the importation of rolled-steel products did not fall substantially below the level of the 1920s; thus we see that rolled-steel imports as a proportion of total rolled-steel consumption fell only from 90 percent in 1930 to about 70 percent in 1940.

It should also be noted that the world depression of the 1930s had an effect on iron and steel consumption for only a short period. It will be clear from Tables 15 and 16 that production levels of pig iron, steel ingot, and rolled products were already higher in 1933 than the previous peak. One should also note that total industrial output had almost entirely recovered by that time.

The generally rapid industrial recovery resulted from the great fall of foreign exchange earnings, which forced Brazil to curtail imports drastically. The resulting shortages of imported manufactured goods and the

31. Edmundo de Macedo Soares e Silva, "Política Metalúrgica Do Brasil," *ABM*, Boletim Da Associação Brasileira de Metais, Janeiro 1946, II No. 2, 14; also his comments in a symposium "Expansão da Siderurgia No Brasil," *Geologia e Metalurgia*, No. 20 (1959), p. 8.

32. It would seem from the data that by the early thirties Brazil was entirely self-sufficient in the production of its pig iron needs and that by 1940 the country was close to self-sufficiency in ingot steel production. This, however, is an illusion. A detailed analysis of the imports of rolled steel products (which is not done here) would show that a large proportion of these products were of a semifinished nature— products destined for rerolling operations in Brazil. The necessity for importing semifinished steel products would thus indicate that Brazil had not become completely self-sufficient in pig iron or ingot steel production.

Historical Developments

61

Table 15

Brazilian Pig Iron, Steel, and Rolled-Steel Production and Consumption, 1916–40

(tons)

Year	Pig iron			Steel ingot			Rolled-steel products		
	Production	Consumption	Imports/ Consumption (%)	Production	Consumption	Imports/ Consumption (%)	Production	Consumption	Imports as a percentage of consumption
1916	4,267								
1917	7,648								
1918	11,748								
1919	10,808								
1920	14,056	21,050	33.2	—	14,409	100.0			
1921	17,747	18,826	5.7	—	7,281	100.0			
1922	17,783	20,839	14.7	—	5,748	100.0			
1923	25,187	30,443	17.3	—	4,276	100.0			
1924	25,035	41,528	39.7	4,492	11,697	61.6			
1925	30,046	41,760	28.0	7,559	14,123	46.5	283	373,485	99.9
1926	21,299	37,530	43.2	9,875	15,908	37.9	10,501	399,381	96.0
1927	15,353	17,618	12.9	8,205	14,581	43.7	16,638	435,767	96.2
1928	25,761	28,356	9.1	21,390	28,932	26.1	26,227	483,149	94.6
1929	33,707	38,129	11.6	26,842	35,712	24.8	29,898	514,296	94.2
1930	35,305	37,258	5.2	20,985	24,766	15.3	25,895	259,224	90.0
1931	28,114	28,756	2.2	23,130	24,777	6.6	18,892	143,489	87.1
1932	28,809	29,786	3.3	34,192	36,431	6.1	29,547	165,650	82.2
1933	46,774	47,975	2.5	53,567	56,469	5.1	42,369	277,028	84.7
1934	58,559	59,282	1.2	61,675	65,904	6.4	48,699	343,590	85.8
1935	64,082	64,581	0.8	64,231	69,390	7.4	52,358	345,389	84.8
1936	78,419	79,670	1.6	73,667	80,426	8.4	62,946	386,689	83.7
1937	98,101	99,313	1.2	76,430	85,746	10.9	71,419	505,352	85.9
1938	122,352	123,210	0.7	92,420	101,433	8.9	85,666	355,662	75.9
1939	160,016	160,374	0.2	114,095	120,842	5.6	100,996	429,845	76.5
1940	185,570	185,999	—	141,201	147,810	4.5	135,293	414,519	69.4

Source: Falcão/CEPAL steel study, op. cit.; IBGE, O Brasil em Numeros; import data for iron and steel ingot imports from Fundação Getúlio Vargas, IBRE.

Table 16

Indexes of Production and Consumption of Pig Iron, Steel Ingots, Rolled-Steel Products, and Industrial Production (1949 = 100)

Year	Pig iron Production	Pig iron Consumption	Ingot steel Production	Ingot steel Consumption	Rolled-steel products Production	Rolled-steel products Consumption	Industrial production
1916	0.8						
1917	1.5						
1918	2.3						
1919	2.1						
1920	2.7	4.1		2.3			
1921	3.5	3.7		1.2			
1922	3.5	4.1		0.9			
1923	4.9	5.9		0.7			
1924	4.9	8.1	0.7	1.9			
1925	5.9	8.2	1.2	2.2		53.5	26.3
1926	4.2	7.3	1.6	2.5	2.2	57.2	27.8
1927	3.0	3.4	1.3	2.3	3.6	62.4	30.4
1928	5.0	5.5	3.5	4.6	5.6	69.2	34.0
1929	6.6	7.4	4.4	5.7	6.4	73.7	32.5
1930	6.9	7.3	3.4	4.0	5.5	37.1	31.4
1931	5.5	5.6	3.8	4.0	4.1	20.5	30.4
1932	5.6	5.8	5.6	5.8	6.3	23.7	30.9
1933	9.1	9.4	8.7	9.0	9.1	39.7	33.5
1934	11.4	11.6	10.0	10.5	10.5	49.2	36.1
1935	12.5	12.6	10.4	11.1	11.2	49.5	39.2
1936	15.3	15.6	12.0	12.9	13.5	55.4	44.8
1937	19.2	19.4	12.4	13.7	15.3	72.4	46.4
1938	23.9	24.0	15.0	16.2	18.4	50.9	49.0
1939	31.3	31.3	18.5	19.3	21.7	61.6	51.5
1940	36.3	36.3	23.0	23.6	29.1	59.4	52.6

Source: Computed from sources of Table 14.

rise of their relative price made it profitable to expand domestic industrial production in substitution for previously imported products. These events explain the rapid recovery of industrial production and the continued expansion noted throughout the rest of the decade.[33]

Most notable in the expansion in the iron and steel industry during the thirties was the building by Belgo-Mineira of its Monlevade plant. With the extension of rail facilities to that area, the company began construction in 1935. In July 1937, its new blast furnace started to operate; in April 1938, the first steel of Monlevade flowed from its new SM furnace; and with the beginning of the operation of its new rolling mill in January 1940, Belgo-Mineira became South America's largest integrated steel works. It also became, and still is, the world's largest charcoal-based integrated steel works. The overwhelming dominance of Belgo-Mineira becomes clear when one compares some of the production data of the firm with total national output. In 1936, with total pig iron production of Brazil equal to 78,419 tons, Belgo-Mineira with its Sabara plant produced 29,518 tons; in 1940, with national pig iron output at 135,293 tons, Belgo-Mineira's output at its Sabara and Monlevade plants amounted to 84,655 tons. The national steel ingot output in 1936 was 73,667 tons, and Belgo-Mineira's Sabara plant produced 30,811 tons of this; in 1940 national output of ingot steel stood at 141,201 tons, while the Sabara and Monlevade plants together produced 85,655 tons. The dominance of Belgo-Mineira was even more pronounced in the production of rolled products. In 1940, Monlevade alone produced 95,556 tons out of a domestic output of 135,293 tons.[34] Table 17 gives a summary view of output of all firms producing in 1939.

The expansion of Belgo-Mineira overshadowed the entire Brazilian steel industry. Throughout the thirties, however, older, smaller firms expanded and a number of newer ones were created. In 1931, the Companhia Ferro-Brasileira was founded in Caeté, Minas Gerais, mainly with

33. See Baer, *op. cit.,* pp. 20–26; there has been and still is some controversy about the nature of industrial growth in the thirties. Some say that shortages of imported manufactured goods stimulated substantial investment in new capacity, while others maintain that much of the growth was a result of the utilization of capacity which previously stood idle. This question has not yet fully been resolved. In the iron and steel industry, both the fuller utilization of existing capacity and the creation of new capacity occurred. See also Carlos M. Pelaez, "A Balança Comercial, a Grande Depressão e a Industrialização Brasileira," *Revista Brasileira de Economia,* Março 1968.

34. The data of Belgo-Mineira's production in 1936 and 1940 were supplied to the author by the firm.

French capital; the firm became the leading producer of iron tubes, using the process of centrifugation; its pig iron output by 1940 had grown to 28,225 tons. Other firms founded in the thirties were Electro Aço Altona S.A., started in 1934 in Blumenau, S. Catarina, a small firm with electric furnaces producing steel castings, small rolled products, mainly bars, and a small amount of forgings; the Siderúrgica Barra Mansa in the state of Rio de Janeiro (close to the present Volta Redonda works) was founded in 1937 with a ten-ton-per-day charcoal blast furnace and has grown ever since, producing medium and light bars, shapes, and wire products; the Companhia Metalúrgica Barbará, founded in 1937 with French and some Brazilian capital, producing mainly centrifugated iron tubes and some iron-foundry products, also in the state of Rio de Janeiro in the region of Barra Mansa. In 1939, the firm Aços Villares was founded to provide castings for its parent company, Elevadores Atlas, S.A. (now the firm Indústrias Villares); the company is wholly owned by the Villares family and the steel firm gradually developed into one of Brazil's leading special steel products producers. It is thus clear that the growth of the Brazilian iron and steel industry in the thirties was based, not only on the growth of already existing capacity, but also on the creation of a substantial amount of new capacity.

Despite these developments, Brazil was still a relatively small steel-producing country by the end of the thirties and, as mentioned previously, in 1940 about 70 percent of rolled steel products consumed were still imported. Approximately 75 percent of all light sections and bars were produced domestically, while all rails and flat products were imported. Given the country's natural resources, it is no wonder that by the latter part of the 1930s Getúlio Vargas and his entourage, especially the military, were anxious to promote the construction of an integrated steel works on a substantially larger scale than Belgo-Mineira, which, with all other smaller producers, could not provide for the increasing consumption of heavy steel products on which the railroads, shipping, and even a large part of the construction industry depended. Vargas thus felt that either private or state, foreign and/or domestic capital should be encouraged to establish a large-scale integrated steel works. Concern for the creation of such an enterprise, however, went back to the early part of the century.

The Itabira Contract Affair

It was mentioned earlier that before the First World War the Brazilian

Brazil: Steel Industry in 1939

Firms	Year of establishment	Number employed	Production in 1939 (tons)		
			Pig iron	Steel	Rolled iron
Cia. Sid. Belgo Mineira S.A.	1921	2,461	72,452	59,155	40,787
Cia. Brasileira de Usinas Metalurgicas					
Morro Grande Plant	1925	681	27,405	745	—
Neves Plant	1925	957	—	21,923	19,487
Cia. Ferro Brasileiro	1931	952	19,235	—	—
Cia. Metalúrgica Barbará	1937	754	8,140	—	—
Cia. Bras. de Mineração e Metalurgia	1925	805	—	28,204	20,907
Cia. Nacional de Navegação Costeira	1938	1,800[a]	—	597	—
Usina Santa Olimpia Limitada	1925	247	—	720	7,167
U. Siderúrgica e Laminadora N.S.Ap.	1938	201	—	—	4,712
Usinas Santa Luzia	1932	325	—	122	—
Laminação e Artefatos de Ferro S.A.	1939	38	—	—	—
Comercio e Industria Souza Noschese	1938	643	2,457	—	—
Pirie, Vilares & Companhia Ltda.	1939	500	1,144	—	—
Siderúrgica Rio Grandense S.A.	1938	134	—	—	2,469
S.A. Metalúrgica Santo Antonio	1931	430	2,816	—	—
Cia. Industrial de Ferro S.A.	1937	145	2,436	—	—
Sociedade Paulista de Ferro Limitada	1936	90	—	95	—
Fabrica de Aço Paulista S.A.	1923	348	—	1,633	—
Usina Metalúrgica Itaite S.A.	1938	78	—	—	829
Usina Queiroz Junior Limitada	1891	412	15,395	—	—
Usina Siderúrgica de Gage Ltda.	1921	81	3,864	—	—
Siderúrgica Barra Mansa S.A.	1937	84	4,673	—	—
Metalúrgica Nestor de Goes Ltda.	1935	96	—	—	2,291
J. L. Aliperti & Irmãos	1924	168	—	360	2,347
Usina Siderúrgica Capiruzinho	1939	—	—	—	—
Eletro Aço Altona Limitada	1936	200	—	541	—
Laminação de Ferro Sacoman Ltda.	1939	14	—	—	—
Total		12,644	160,017	114,095	100,996

Source: José Jobim, *The Mineral Wealth of Brazil* (Livraria José Olympio, Rio de Janeiro, 1942), p. 42.

[a] Includes laborers in activities other than iron and steel.

government had shown an interest in promoting a large-scale, integrated steel works, financed by foreign capital. Shortly after the war the idea of a large-scale, integrated steel works was revived by an American entrepreneur of long experience in Latin America and especially Brazil, Percival Farquhar. Farquhar had been active in organizing a variety of ventures since the late 1890s, ranging from the building of the Cuban Railway, the Guatemalan Railway, the Rio de Janeiro Tramway, the Rio Light & Power Company, the Port of Pará, the Madeira-Mamoré Railway in the interior of Brazil, to the promotion of cattle-raising and lumber operations in Southern Brazil.[35] In late 1919 and early 1920, Farquhar presented a plan to the President of Brazil, Epitácio Pessôa, combining the export of iron ore with the construction of a large-scale, integrated steel enterprise. Although after many years of struggle the plan never became a reality, it is of interest to summarize the episode because it is the best example of the noneconomic difficulties Brazil faced in starting to produce steel on a large scale. There, noneconomic factors consisted mainly of the crosscurrents of political and nationalist forces. The episode is also of interest because the basic ideas of Farquhar were ultimately put into operation in the post–World War II period.

In 1911, an English capitalist, Sir Ernest Cassel, founded the Itabira Iron Ore Company, which owned Cauê Peak in Minas Gerais (also called the "iron mountain") and 18,000 acres around it. After a world congress on iron resources in Stockholm in 1910, where there had been reports on the abundance and purity of Brazilian iron ore, there was a general rush by foreign speculators to buy iron ore lands in Minas Gerais. Many Brazilians, especially state and federal government officials, were resentful over these land acquisitions and did everything in their power to prevent the export of the iron ore. They also had the erroneous notion that Brazil was sitting on a large proportion of the world's iron ore reserves and that by staying out of the world market, Brazil would sooner or later force up world iron ore prices, making iron export substantially more lucrative. They did not count on the equally rich iron ore discoveries in Canada and Venezuela. Thus, the Itabira company, not having the necessary government contract, could not commence operations. In 1918 Cassel sold the enterprise to a group of British bankers and steelmakers, who after a while were looking for an American buyer.[36]

35. The only existing biography of this entrepreneur (1864–1953) is Charles A. Gauld, *The Last Titan: Percival Farquhar, American Entrepreneur in Latin America.*
36. Gauld, *op. cit.,* pp. 282–284.

Farquhar, with financial backing from a number of American business interests, acquired control of the company. In July 1919, when President-elect Pessôa came through New York on his way to the Versailles Peace Conference, Farquhar found him receptive to his ideas. He proposed an authorization for Itabira Iron to export four million tons of iron ore a year to build an iron ore port at Santa Cruz (56 kms. north of Vitoria), to build a new broad-gauge railway parallel to the poorly constructed existing one (the Estrada de Ferro Vitória a Minas), and to build an integrated steel mill in the area. The idea was to export iron ore over the new railroad and port and to bring back from Europe or the United States coal in the empty ore ships. The site of the proposed steel works of 150,000 tons was also the new port of Santa Cruz. In January 1920, the Itabira Contract was passed by the Brazilian Congress. For political reasons, President Pessôa made the contract conditional on the approval of the legislature of the state of Minas Gerais, which ultimately proved fatal to the undertaking.

Farquhar's initial difficulties after approval of the contract were with the political forces of Minas Gerais, especially the nationalistic Governor Artur Bernardes, whose approval was needed. Among the reasons for Bernardes's opposition was his conviction that Farquhar had no serious intention of combining his ore-exporting scheme with a large steel plant. Some writers have claimed that in fact Farquhar tried to escape the steel commitment, since he was primarily interested in the iron ore–exporting business. He made a determined effort, however, to get European and American financing for the steel works. He found foreign financial interests connected with steel to be concerned mainly with obtaining Brazilian iron ore and manganese for European and American steel plants. There was little interest in constructing new steel works in an underdeveloped region. In short, the financial interests willing to back Farquhar's iron ore–exporting scheme could not be persuaded to raise money for the steel works.[37]

In 1928, Farquhar gladly surrendered a number of monopoly privileges acquired in the 1920 contract (monopoly transit on the new railroad, monopoly port privileges) in order to be relieved of the steel plant obligation. In 1929, the Itabira venture almost succeeded, but the crash and the following world depression closed the financial markets for such a project.

37. Wirth, *op. cit.*, pp. 100–101.

In the first years of the Getúlio Vargas administration, until the Estado Novo (1937), the Itabira Contract would have had no political difficulty in getting under way. With the establishment of the Estado Novo, however, the Brazilian government again insisted on combining the iron ore–exporting scheme with the erection of a steel plant. The only hope for Farquhar after 1937 was to use his connections with the German steelman Fritz Thyssen. It seems that plans had been considered for an exchange of German railroad and steel equipment for yearly shipments of four million tons of Itabira iron ore.[38] Given the world situation in 1938–39, however, and the Brazilian foreign minister Aranha's pro-American feelings, this plan was blocked by the government. In August 1939, Vargas finally decreed the espiration of the Itabira Contract.

Farquhar did not give up at this stage. With some Brazilian associates, he founded the Companhia Brasileira de Mineração e Siderurgia which would buy Farquhar's Itabira option without the steel mill. In September 1939, Vargas authorized the new company and gave it the right to mine and export iron ore and to operate the Vitória a Minas railroad. In the following years, however, Farquhar faced difficulties in obtaining financial backing for the company. In 1942 the Brazilian government finally expropriated the enterprise. With the money paid him by the Brazilian government, Farquhar and some Brazilian associates founded the steel firm ACESITA (Aços Especiais Itabira) which was built near the site of Itabira. Farquhar justified the site of the mill, a long distance from consuming centers, on the grounds that the firm would be producing special steels with quality so high that in spite of transportation costs they would still be competitive. As the steel mill was erected, more and more financing had to be obtained from the Banco do Brasil, and by 1952 the bank had full control of the company.

Towards a National Steel Solution

Throughout the decade of the thirties, Brazil's political, military and economic leaders searched for ways of promoting the erection of a large and integrated steel plant. In 1931, Getúlio Vargas said in a speech in Belo Horizonte, "The most basic problem of our economy is that of steel. For Brazil, the age of iron will signify its economic opulence."[39] This view was also firmly held by Brazil's military establishment. Accordingly, in 1931, the Minister of War, Leite de Castro, created a National Steel

38. Wirth, *op. cit.*, p. 102.
39. Quoted in Bastos, *op. cit.*, p. 131.

Commission in order to study the whole steel problem.[40] This commission served as a focus of study and debate throughout the thirties, and from its deliberations and through the activities of many of its members resulted the ultimate creation of the Companhia Siderúrgica Nacional, which built Volta Redonda.

The reason for the increasing government interest in building a steel mill was largely that private capital (especially foreign capital) for large scale steel works was not forthcoming (it should be remembered that Farquhar at times tried to avoid the steel mill commitment and at other times had difficulties finding interested parties in the United States or Europe). Existing steel enterprises were too small and had no funds for sizable expansion programs, and Belgo-Mineira could not or was not prepared to expand beyond the efforts it had made in constructing the Monlevade plant. The erection of the latter was not enough to keep pace with the country's steel needs. It did not have the capacity for producing heavier steels for the shipping industry, rails, and products which the construction industry needed. It was also felt that a real breakthrough in the substitution of heavy steel imports could only be done through the building of a large-scale coke-using plant. The country's forest reserves were not thought adequate to supply a really large-scale charcoal-based steel industry.[41]

When during the thirties the Vargas regime was still giving serious consideration to the Itabira project, the opposition to such a project was also formed by owners of the existing steel industry. They feared that the construction of large-scale steel works would imply their ruin. A leader of this opposition was Alexandre Siciliano, a planter from São Paulo who also had interests in metallurgical plants based on scrap. He argued that not only would a large-scale mill based on coke ruin the existing industry, but that it would put into jeopardy the military requirements of economic self-sufficiency, because of the country's dependence on imported coal for coke-based operations. He therefore proposed the establishment of a domestic steel industry organized by the government.

Siciliano proposed the erection of a 150,000-ton steel plant at Juiz de

40. A member of this commission was lieutenant Macedo Soares e Silva, who was to be an important figure in the development of the Brazilian steel industry. After tenente revolt in 1922–24, he escaped to France and studied metallurgy. After the 1930 revolution, he served in the São Paulo government and later in the decade was one of the principal leaders in the establishment of Volta Redonda, where he eventually became president.

41. Bastos, *op. cit.*, p. 157.

Fora (Minas Gerais), which would produce ingots for privately owned rolling mills in Rio de Janeiro, São Paulo, and Minas Gerais. His plan included the erection of a separate coke plant in the city of Entre Rios, in the state of Rio de Janeiro, which would first use imported coal, to be replaced gradually by domestic coal; this coke plant would supply the steel mill, would produce gas to illuminate neighboring cities, and would produce byproducts for the production of explosives. Siciliano felt that already existing blast furnaces would not be disadvantaged by his plan and suggested that they could produce high-grade pig iron for special steels. The main objection to the plan was that modern steel technology dictated the complete integration of a large-scale steel mill, not the separation of such important units as the blast furnace from the coke oven. Also, Siciliano's location made the mill dependent on the services of the inefficient Central do Brasil railroad.

Some other plans were put forth during the middle thirties, but all were poorly conceived.[42] The government showed itself indecisive. Pressure increased on Vargas, however, especially from the armed forces, to come up with a definite plan for a large-scale steel breakthrough. In February 1938, Vargas declared that he considered three possible approaches to the establishment of a large steel mill: a state-run enterprise with foreign financing (or financing by iron ore exports); a mixed Brazilian company with joint government and private participation; or a private company with foreign participation, but under state control or supervision. Vargas once more called for a simultaneous development of large-scale iron ore exports and the establishment of an integrated steel mill.

It also seems that by the late thirties the steel establishments in the principal industrial centers of the world regarded the possibilities of a large-scale Brazilian steel industry with a greater degree of seriousness than previously. Until then, no one wanted to disturb the established markets, and it was generally felt that it was "unnatural" to develop a steel industry in a steel-importing nation—that it was doubtful that Brazil had a comparative advantage in such an enterprise. A change in attitude came about partially due to a realization that Brazil had a large and growing internal market for steel products (the possibilities of such a market probably became clearer during the industrial growth which oc-

42. For example, Henrique Lage, the owner of coal mines in Santa Catarina and also a large shipowner, tried on various occasions to interest the government in developing steel firms using his coal; the plans usually included substantial subsidies to help modernize his coal operations. See Wirth, *op. cit.,* pp. 110–111.

curred in the thirties), and that with such a market and the large iron ore reserves the development of the steel industry would make sense. The changed attitude was also partially influenced by the political prestige for a country associated with helping Brazil build up such an industry.

Both Germany and the United States were interested in the possibility of contributing to the erection of a large-scale steel mill. Two groups developed within Brazil. One was composed of nationalists and some parts of the military establishment who wanted to export iron ore to Germany in return for equipment to erect a steel mill and to expand and modernize the railroad system, needed to complement the growth of the steel industry. The second group was led mainly by Brazil's diplomatic establishment which, given the darkening world political situation in the late thirties, feared German economic and political penetration into Brazil and therefore wanted United States capital to finance a steel enterprise. The leader of this group was Oswaldo Aranha, Brazil's ambassador to Washington (1934–37) and later foreign minister. Aranha had the political support of Vargas.

A study made in 1937 by American engineers for the DuPont enterprises, which were interested in the possibility of investing in Brazil, gave some optimistic results. It was estimated that in 1937 a 200,000-ton capacity steel mill could substitute US$48 million worth of imports. It was shown that the country needed 75,000 tons of rails and much additional steel for construction, plates for shipyards, and tinplate for a national canning industry. The engineers recommended that a tidewater plant be located either in Vitoria or Rio de Janeiro. To the surprise of everyone, they felt that the Central do Brasil Railroad, with some improvements, could support a large steel mill. They declared national coal to be unsuitable and recommended the construction of a stand-by charcoal-based blast furnace in case the exterior coal supply should be cut off. Costs at that time were estimated at US$18 million, of which US$10 million represented imported equipment and the rest local funds for construction. Thus, the participation of Brazilian capital was strongly recommended. Intense nationalistic pressures on Vargas at the time, however, favored establishing a state-owned plant, and the DuPont project fell through.

In 1937–38, some influencial Brazilians sought to have the government expropriate some of the best iron ore mines (the Casa da Pedra, for example, controlled by the German firm A. Thun & Co.) and to have the government export iron ore in return for equipment for a large-scale steel mill. Early in 1938, the German steel firm Demag was approached by the

Brazilian government. It was proposed that in return for one million tons of iron ore a year the firm would build a steel plant, rail, port, and mining installations. Although the firm and the German government were interested for both economic and political reasons, the priority was not high enough for them to accept the Brazilian terms outright. The German government refused to become a direct partner of the enterprise (this had been initially requested by the Brazilians) and it accepted only 50 percent of the payment in iron ore, while the other half would have to be paid in raw materials of the Germans' choosing on a short-term basis.

During 1938, Demag created an informal consortium with Stahlunion (a subsidiary of Vereinigte Stahlwerke, which had exported some iron ore in 1937 from its Brazilian holdings) and Krupp, who had already declared himself willing to build steel shops for the Brazilian army in a separate agreement. This consortium broke up very soon, however, with the withdrawal of Stahlunion. Fritz Thyssen, the head of Vereinigte Stahlwerke, had been working independently with Farquhar, which possibly explains this move.

The jockeying for positions continued; meanwhile the debate by the Steel Commission over the location of a steelworks went on, the majority finally favoring a plant in or near Rio. Vargas at the time recognized the army as being the final arbiter of the entire steel problem.

In January 1939, Macedo Soares went to Europe to investigate the possibilities of one or a group of German firms setting up the steel mill. He talked to the firms Gutehoffnungshuette, Stahlunion, and Krupp, and even investigated the possibility of British financing. At the same time, Oswaldo Aranha, the partisan of a United States collaboration, had gone to the United States late in 1938 to talk about a wide range of political and economic topics with the American government. Among the things discussed was the possibility of United States Government credits for the development of Brazilian raw material exports and a steel plant. Aranha probably stressed the fact that if United States financing were not available, Brazil would have to turn to Germany. This argument had its intended effect, and the Export-Import Bank made a tentative commitment to finance the acquisition of equipment for a Brazil iron and steel plant. With this commitment, Macedo Soares was ordered to break off his negotiations in Europe and sail for the United States.

Aranha, after trying unsuccessfully to interest the DuPonts to participate again, turned to the United States Steel Corporation (USS). USS had an interest in looking into the possibilities of furnishing equipment for the development of Brazilian iron ore mining. The company was against a state-owned steel plant, however, and the expensive railroad construction in the Rio Doce valley. It suggested a joint American-Brazilian company to operate a steel plant. USS, impressed also by the older DuPont report, sent a technical mission to Brazil under the leadership of the president of its export subsidiary, H. Greenwood.

The USS mission arrived in Brazil in June 1939 and studied possible plant locations, the Santa Catarina coal fields, iron ore resources, transport and market conditions. Macedo Soares was head of a group of Brazilian experts who co-operated with the mission, and in October a joint report was issued. The mission found Brazilian coal usable for blast furnace operations (though it recommended first using imported coal exclusively, gradually mixing it with national coal until an optimal mixture was reached). It also recommended that the steel mill depend on the Central do Brasil rather than the Rio Doce Railroad (it was found that the Central could carry enough ore for a large mill and enough export ore to compensate for the large coal imports.[43] It suggested that the best site was one near Rio de Janeiro, at Santa Cruz on the Sepetiba Bay, where sea and land transport would meet halfway between the coal fields in the South and the ore mines of Minas Gerais. The site was market-oriented and had abundant water supplies. The production program envisioned included rails, shapes, plates, sheets, and tin plate. This program was selected with a view of not interfering with existing domestic products.

The solution was ideal, since it met all requirements. It met the requirements of a feasible large-scale steel operation, it conciliated the interests of existing producers, it met military requirements, and it provided for improved transportation routes. Also, this proposal was not directly tied to iron ore exports. It seemed in fact so superior to all previous proposals that it was one of the contributory factors in the final cancellation of the Itabira Contract.

The proposal put forth was to form a Brazilian company to be controlled by U. S. Steel. This control would be established through the common stock it would obtain in return for US$5 million worth of equipment. Brazilian private capital would participate (especially Heitor de Carvalho of the Paulista railroad and the powerful Guinle family), but in a

43. This solution was also satisfying from the point of view of the military, who were mainly concerned about potential self-sufficiency, since in case of war, continuous, though expensive, coal supplies would be assured.

minority capacity. The Export-Import Bank was supposed to make substantial loans for equipment purchases.

President Vargas and his immediate circle were most enthusiastic about the proposals of the Greenwood Mission, and the mission returned to the United States for approval of the U. S. Steel company's finance committee.

Trouble started even before the Greenwood Mission presented its report. Brazilian nationalist forces, especially in the army, were dissatisfied with the idea of foreign financing and control of a large-scale steel mill, and they were also opposed to the mining operations of U. S. Steel. These nationalist forces succeeded in passing a Mining Code in January 1940 which prohibited foreign capital in mining and metallurgy. This occurred at the moment that the finance committee of U. S. Steel was meeting to decide whether to accept the proposed participation in the Brazilian steel project.

U. S. Steel made it clear to the Brazilian government that its participation depended on a modification of Brazilian laws, in order to enable it to control the steel works, to protect the company's subsoil rights on its mining properties, and to enable United States technicians to work permanently in the steel operations. The issues at stake are well summarized by Wirth when he states that the "nub of the conflict centered on the final purpose of the projected steel works. Clearly, the nationalists wanted to preserve all development in the metallurgical and mining sectors for Brazilians. On the other hand, the U. S. Steel Corporation was attracted by the growing South American steel market and it wanted to establish a Brazilian base of operations."[44]

As 1939 drew to a close, Vargas and his ministers became increasingly conciliatory, indicating their willingness to discuss the report of the commission, and it seems probable that most of the commission's recommendations would have been accepted. It was thus a shock for Vargas and his cabinet to learn in the middle of January 1940 that the U. S. Steel Corporation had decided to abandon the Brazilian steel project. Several explanations exist for the company's exit from the project. It was claimed that the project would require an investment of US$70 million—too much of a burden on the corporation at the time, in view of the fact that no dividends had been distributed for the four previous years. Also, since the corporation had lost its nickel properties when

44. Wirth, *op. cit.*, p. 143

Russia invaded Finland, the company at that time had fears about the dangers of investing abroad. It was also claimed that there was a conflict within the corporation's administration between those interested in Brazil's economic potentials and those fearful of losing current export markets.[45]

The company's decision was a big blow to the Vargas government. Vargas soon put substantial pressure on the United States Government "to do something in the light of the Good Neighbor Policy." It seems that Vargas was anxious to have the steel project undertaken, if possible, by the collaboration of private domestic and foreign interests. The Brazilian government and the United States State Department tried a number of times to get the U. S. Steel Company back into the picture, but without success.

During 1940, a number of steel companies and engineering firms were invited to look into the Brazilian steel problem. Among these were the firms of Brassert, Arthur G. McKee, and the Swedish armaments and steel firm Wenner-Gren. Even Krupp was in the picture, described by contemporary United States and Brazilian press reports as eager to finance and construct the steel works. Vargas and his foreign minister Aranha, however, were still anxious to rely, if possible, on United States help.

To achieve his aims, Vargas in the early part of 1940 made a number of moves towards the formation of a steel company and towards obtaining American financial help. He organized an executive commission in March 1940 under the leadership of the industrialist Guilherme Guinle to prepare the definitive national steel plan. A prominent member of the commission was Macedo Soares, the steel expert, who would be the man responsible for the technical execution of the project. Initial financing was arranged from a variety of sources: the government savings bank, pension funds, and private sources (businessmen and bankers).

As the first half of 1940 drew to a close, Vargas's hope for a private-enterprise steel works was fading. Vargas then decided to take a political gamble which ultimately paid off.[46] He made it clear to the United States that he considered the steel project the key to Brazilian economic development and a test of Brazilian–United States co-operation under the Good Neighbor Policy. Vargas also invited the Germans to make an offer. This was obviously a tactic to place additional pressure on Washington.[47]

45. *Ibid.*, p. 144; also Bastos. *op. cit.*, pp. 186–187.
46. See Wirth, *op. cit.*, pp. 147–148; Bastos, *op. cit.*, pp. 187–189.
47. The fact that this tactic influenced Washington can be gathered from pub-

The United States Government at this stage indicated that it wanted to discuss its support for the steel project in conjunction with certain military and political considerations, especially the stationing of United States troops at strategic Brazilian points in case of war.

The Brazilian steel commission arrived in Washington in July 1940, shortly before President Roosevelt decided to give Brazil and Mexico first priority on military and economic aid. In September 1940, the Export-Import Bank pledged US$20 million to finance the steel works. It seems that the Germans were outbid, since they were not in a position to guarantee immediate delivery of equipments; their terms were less favorable than those of the United States, and they demanded that half of the short-term payments in raw materials be in German hands before delivery of equipment.

The Building of Volta Redonda

On the 9th of April 1941, the Companhia Siderúrgica Nacional (CSN) was founded. The savings banks and pension funds supplied about half of the initial capital of US$25 million. Guinle, its president, tried hard to interest private capital to participate but had only moderate success. The Brazilian treasury purchased all the common stock not subscribed to by the public. When, later on, domestic capital was increased, private capital made only very small contributions and the control of the government increased even more. For all practical purposes CSN became a government-owned and operated company.

The location of the steel works was decided upon before the official founding of the company. In July 1940, the Steel Commission had selected the Volta Redonda site, situated in the state of Rio de Janeiro, approximately 100 kms. from the city of Rio de Janeiro and about 450 kms. from São Paulo. A number of factors contributed to the selection of the site. It was claimed that Volta Redonda was a place where raw materials could be assembled at a relatively low cost, and from which the finished

---

lished U. S. diplomatic papers of that period. In a letter to the Federal Loan Administrator Jesse Jones, Sumner Welles, the Acting Secretary of State, writes that "failure on the part of . . . [our] Government to assist the Brazilians in this matter will in all probability according to the American Ambassador to Brazil result in the immediate acceptance by Brazil of a German offer to build a plant, which the Germans are prepared to do on terms which they will allow the Brazilions to write themselves. Germany's predominance in Brazilian economic and military life would thereby be assured for many years." *Foreign Relations of the United States, Diplomatic Papers, 1940,* V, "The American Republics," pp. 609–610.

steel products could be easily shipped to the two principal consuming markets. The sites near Rio de Janeiro (Santa Cruz) or at Vitoria were discounted in order to avoid the expenses of constructing port facilities and entirely new railroad links and also for military considerations, such as vulnerability to enemy attack from the sea. It was also claimed that a place about 1,500 feet above sea level provided generally healthier working conditions. Also taken into account was Volta Redonda's location in the economically depressed Paraiba Valley, where wages at the time were about 30 percent lower than in Rio de Janeiro. There can also be no doubt that political as well as economic considerations entered into the site selection. The military were sensitive to the security arguments and thus favorable to Volta Redonda; the site was also a compromise between the power centers of Rio, São Paulo, and Minas Gerais; finally, the fact that Vargas' son-in-law Amaral Peixoto was an influential figure in the government of the state of Rio de Janeiro probably had also some influence in the decision.[48]

Many of the locational advantages claimed for the Volta Redonda site at the time of the selection were open to doubt a few years after the plant was built. For example, the transport system which seemed adequate in the early forties did not keep up with the postwar industrial growth and created severe problems for Volta Redonda. Also, the cheap labor argument was counter-balanced by the necessity for building a large industrial town, which would have been unnecessary had the site selected been near a large city. Finally, it turned out that the cheaper labor argument was offset by the fact that labor costs in relation to other costs is rather small and that much of the labor at hand had to receive special training, since most available workers had never before worked with machinery. In the early years of Volta Redonda the accident rate was extremely high.

48. The best official defense for the selection of the site was given by Guinle, the head of the Steel Commission, who stated, "The best location would be the Federal District [Rio de Janeiro at the time]. But this site would imply the construction of a port at Santa Cruz at a high cost, would imply substantial drainage operations, the construction of a pipeline to supply the plant with fresh water . . . and the reconstruction of 40 kms of railroad track. Besides this the Santa Cruz site would imply substantial expenditures to construct military installations to defend the plant.

"In choosing a site in the Paraiba Valley . . . the commission considered that at 400 meters altitude the working conditions would be better, that the price of labor there was cheaper than in Rio de Janeiro, that there was an abundance of fresh water there and that the Central do Brasil railroad would thus get a good return from freights paid by the company." Freely translated from quotation in Bastos, *op. cit.,* p. 230.

In September 1940, the United States Export-Import Bank lent US\$ 20 million for the project. The only condition imposed was that, given Brazil's inexperience in the production of steel on a large-scale basis the management of the firm should include engineers and administrators from the United States with experience in steel until to the mutual satisfaction of both parties the firm could be run entirely by Brazilian technicians. As construction proceeded, the loan was to rise to US\$45 million because of costs higher than anticipated, resulting mainly from rising prices in wartime conditions.

Until the end of 1941, Macedo Soares remained in the United States to purchase equipment—turbo-generators, coke oven equipment, blast furnace equipment, equipment for the steel furnace, the rolling mills, electrical equipment. Wartime conditions and the direct controls which the American government imposed on its economy, giving priorities to production efforts which supported the war through the War Production Board, brought the Brazilians frequent difficulties in obtaining various types of equipment on time. One firm, which had won the competition to supply the steel shop equipment, was put on the United States government's blacklist. Because of wartime obligations, the firm Mesta Machine Co. of Pittsburgh had to delay again and again the construction of equipment for the slabbing mill, the key to the entire rolling mill complex. Wartime conditions also increased prices to be paid for equipment by about 60 percent. This was both because of the rise of prices in the United States and the rise of maritime freight prices and insurance. The latter price increases were the direct result of the dangers of maritime shipping during wartime conditions.

In March 1941, the land for the steel works was expropriated, and in late 1941 Macedo Soares returned to Brazil to take charge of the construcion works. By January 1941, the Steel Commission had announced that the American firm Arthur G. McKee & Company had been chosen to make the definitive design of the project; later on, when Guinle had become president of the newly formed Companhia Siderúrgica Nacional, he also announced that Arthur McKee would expedite the purchase of the necessary machinery.

As construction got under way, the work force constantly grew, until at one time 55 American experts, 127 Brazilian engineers, and about 7,000 workers were employed. A large proportion of the Brazilian technical personnel was provided by the armed forces, which, as already mentioned, had for some time been training metallurgical engineers and had already sent many abroad for specialized training. The growth of Brazilian qualified technical personnel was so rapid that by 1947 they were able to assume full control over all operations.

Despite the difficulties of working in wartime conditions, the project advanced steadily, and by the end of 1944 the coke plant, blast furnace, most of the thermoelectric plant, the water treatment plant, repairs shops and some other minor departments were almost completed. The steel shop was about half finished, but the rolling mill section was far behind schedule, mainly because of the delays in delivery. By the end of 1945, the mill was about 80 percent completed. Considering the wartime difficulties, this was a relatively fast construction period, since the site preparation only began in 1942.

In April 1946, Volta Redonda began to produce coke (the first coke production for steel in Brazil), and in June of that year the blast furnace and the steel shop began to function. The rolling mills, however, were not finished until late 1947 and began to operate only in 1948.

POSTWAR EXPANSION OF BRAZIL'S STEEL-PRODUCING CAPACITY

As mentioned in an earlier section of this chapter, throughout the thirties and forties smaller firms were founded, many of a semi-integrated nature (firms producing steel using electric furnaces to melt down scrap). Often these enterprises began as subsidiaries of firms that wanted to integrate backward, producing various types of equipment. There were other examples of steel mills being founded by private entrepreneurs, but slowly becoming government-owned firms through their indebtedness (see the above-mentioned case of ACESITA).

Industrial growth in the post–World War II period, and especially the emphasis in the fifties on the vertical integration of Brazil's industrial complex,[49] led to the expansion of existing steel-producing facilities and the creation of a number of new enterprises, both private and governmental. Volta Redonda, whose ingot capacity was 270,000 tons when it opened in 1946, went through successive expansions until its capacity in 1965 was approximately 1,400,000 ingot tons. Belgo Mineira's successive expansions resulted in a capacity of 450,000 tons in 1965. Many of the smaller private firms, both semi-integrated and integrated, also expanded with the growth of the internal market.

49. For an interpretation of Brazil's industrialization process emphasizing the vertical integration of the industrial complex, see Baer, *op. cit.*, chapters three and six.

A number of new smaller firms made their appearance in the fifties. The largest of these was the German enterprise Mannesmann, which founded its Brazilian subsidiary Companhia Siderúrgica Mannesmann in 1952 and began to build an integrated steel mill specializing in the production of seamless tubes and of medium and heavy bars in special as well as carbon steels. Its plant was built in the industrial city of Belo Horizonte, near an ore mine acquired by the firm, and by 1956 most of the plant's sections were in full operation.

Mannesmann was one of the two firms, the other one being ACESITA, which installed an electric furnace for reducing iron ore into pig iron. It built a coke-using blast furnace only in its expansion in the early sixties. Initially, its steel shop consisted of electric steel furnaces. Only during its expansion in the early sixties was an LD furnace installed. Through various expansion periods, Mannesmann reached production levels of more than 200,000 tons of ingot steel in 1964 and more than 160,000 tons of rolled products.

During the fifties, the erection of two new large integrated steel mills was conceived—Usiminas and Cosipa. The basic initiative for the creation of these firms stemmed from local private and local government interests.

The idea for the creation of an integrated steel complex which was market-oriented in its location—located near São Paulo—was conceived in 1951 by the Paulista engineer Plínio de Queiroz, while visiting Volta Redonda. He was supported from the beginning by some powerful financial interest groups from São Paulo. The Companhia Siderúrgica Paulista (COSIPA) was officially founded in 1953, and Queiroz was its first president. Over the years, as a number of alternative projects were looked into and after two separate projects were worked out (one by the Koppers Company and one by the Kaiser Company), it became obvious that the private groups participating in the company would not be able to come up with all the capital necessary for the realization of the project. In 1956, the state of São Paulo supplied additional capital by becoming a shareholder in Cosipa, and a little later on the Brazilian National Economic Development Bank (BNDE) did likewise, gradually become the majority shareowner of the enterprise. By the mid-sixties the BNDE controlled 58.2 percent of the shares, the state of São Paulo 23.3 percent, the national treasury 6.7 percent, and the balance was owned by various mixed government companies and private groups. It was not until the end of the fifties that construction on Cosipa began. The building of the steelworks took substantially longer than was initially planned and was considerably more expensive than originally estimated because of the nature of the terrain on which the plant was built near Santos (it turned out that the area for the plant site had extremely poor soil conditions). It was only in December 1963 that the rolling mill section began to operate, rolling ingots and slabs from Volta Redonda and Usiminas, and only by the end of 1965 did all the major sections of the mill enter into operation. Cosipa was designed to produce about two million tons of flat products, although the original capacity of its blast furnace and LD steel shop made it possible to reach only production levels of a 600,000 tons. The design and construction of the plant was entrusted to the American engineering firm Kaiser.[50]

The basic idea for the construction of Usiminas also came initially from local private and local government interests. It will be remembered that the dream of establishing an integrated steel plant in the area of the Rio Doce or Paraopeba valleys, using coal coming from abroad in empty iron ore ships and trains, goes back quite a number of years. Farquhar's original scheme of linking iron ore exporting with steel producing was finally realized in the combination of the activities of the government iron-exporting firm Companhia Vale do Rio Doce and the firm Usiminas.

Usiminas was created in 1956. Since it was obvious that the technical co-operation and resources of foreign capital were needed to build the project, negotiations were immediately started with Japanese, German, and even some East European groups. Finally, in 1957, an accord was reached with a Japanese group. The latter, representing various Japanese firms, formed itself into the Nippon-Usiminas Kabushiki Kaisha group, and in exchange for a 40 percent equity participation in Usiminas undertook to plan and supervise the erection of the steel mill and to supply the equipment, most of it coming from Japan. The Development Bank originally participated to the extent of 24.64 percent, the government of the state of Minas Gerais 23.95 percent, the Companhia Vale do Rio Doce 9 percent, the Companhia Siderúrgica Nacional 1.52 percent, and private participation only accounted for about one percent.[51]

Preparation of the site of Usiminas, located in the Vale do Rio Doce

50. "Breve História De Cosipa," IBS, *Boletim do Instituto Brasileiro de Siderurgia,* Ano III, 2 Trimestre (1966), pp. 18–19, "A Cosipa de Ontem, de Hoje, e de Amanhã," by Pedro Dias de Souza, IBS, *Boletim do Instituto Brasileiro de Siderurgia,* Ano V, Março de 1968, pp. 9–27.

51. "O QUE É A USIMINAS, unpublished mimeographed paper of the firm summarizing some basic facts of the enterprise for the visitor.

a few kilometers of the site of Acesita on the Vitoria a Minas railroad (constructed to ship iron ore from the Minas Gerais mines to the port of Vitoria), started in 1958 and construction of the plant began in the following year. Most of the sections of the mill were completed by the end of 1962, and in November of that year the blast furnace began to function. By late 1963 all the main sections of the mill were in operation. In 1966, Usiminas reached production levels of more than 500,000 ingot steel and nearly 390,000 flat rolled products. Since construction costs increased substantially over the construction period, the BNDE was forced to make increasing contributions to the project in order to see it to a successful conclusion. In so doing, the BNDE became the majority stockholder of the firm, and Japanese participation shrank to a little more than 20 percent.

Like Cosipa, Usiminas is an integrated, coke-based steel mill, specializing in flat products. Its plant layout was such that the firm could be able to attain production levels of about two million tons of rolled products. By 1966 only a relatively small investment was needed to expand the blast furnace and LD steel shop section of the mill in order to bring production levels to more than one million tons of rolled products. I mentioned in the previous chapter that a substantial amount of investment funds of Usiminas went into the construction of a new industrial town, since the site of the plant was in an unpopulated region. This type of investment was not necessary for Cosipa, whose location near Santos enabled it to draw its workers from that city.

The erection of Cosipa and Usiminas represented a major expansion of Brazil's capacity to produce flat steel products. The industrial boom of the fifties, however, also stimulated the expansion of many smaller steel firms and the appearance of quite a number of new small firms. In Minas Gerais, a number of small integrated steel establishments were created for the production of various types of nonflat products. The BNDE, which in the fifties had acquired the Companhia Ferro e Aço de Vitória (it was originally a private concern running a blast furnace operation), decided in 1959 to expand that firm by building a rolling mill to reroll blooms from Usiminas, CSN, and other companies into light bars and medium structures. The plant was built by the German firm Ferrostaal A.G., which also became a small shareholder. Ferro e Aço de Vitória began to produce its first rolled products in 1964. As of this writing, there are plans to integrate the firm backward—turn it into a large integrated mill producing nonflat products.

An interesting phenomenon that occurred in the fifties was the boom in small independent blast furnace operations, during a period of substantial pig iron shortage. This stimulated a large number of small entrepreneurs (mainly in the state of Minas Gerais) who owned land with iron ore deposits to build small charcoal-using blast furnaces. Many of these furnaces had a daily capacity of only 20 to 40 tons. By the early sixties, there were 89 such operations in existence. The total yearly capacity of these furnaces was close to one million tons. With the expansion of larger enterprises, however, and the appearance of large new firms producing their own pig iron and even selling it to other firms, these smaller operations were hit by a severe crisis that forced about two-thirds of them to close down.[52]

By 1966, with an ingot production of more than 3.7 million tons, Brazil was the largest steel-producing nation in Latin America. About two thirds of Brazil's steel-producing capacity was in the hands of government-controlled firms. It is interesting to note that in each case the government came in reluctantly. A private solution was first contemplated in the case of Volta Redonda; this failed mainly with the bowing out of U.S. Steel. Cosipa was the brainchild of private groups and Usiminas of local private and local government groups. In each case, the costs turned out to be too much of a burden for any one domestic private group, and since the government in its industrialization policies was determined to encourage the maximum amount of import substitution in basic industries, it made its development bank underwrite the successful conclusion of these projects.[53] One steel expert in Brazil has claimed that given the huge investments necessary for the erection of a flat-products integrated mill, it was impossible for any private Brazilian group to finance such an undertaking. He stated, however, that the smaller scale of operation and subsequent lower capital costs of nonflat-products steel mills made this field of investment much more accessible for private Brazilian groups.[54]

52. See Appendix II for more details on capacity, output and location of these small firms.

53. The growth of government-controlled firms in the steel industry did not preempt the field from private industry. In 1966 a private Brazilian group finished the construction of a new special steels plant (Aços Anhanguera), the same group is toying with the idea of building a large integrated plant for exporting semifinished steel products, and such firms as Belgo-Mineira have substantial expansion plans.

54. Amaro Lanari Junior, "O Projeto da Usiminas e sua Justificatova no Planejamento da Siderurgia Brasileira," *Geologia e Metalurgia,* No. 23 (1961), p. 264.

5

THE GROWTH OF STEEL OUTPUT
AND ITS IMPACT ON THE
BRAZILIAN ECONOMY

THE EXPANSION of the iron and steel industry of Brazil is recorded in Tables 18 to 20. The growth of the industry on a yearly basis since 1940 may be followed in Tables 18 and 19, while Table 20 summarizes the development of the industry since the early part of the century. It would seem from the data that by the early thirties Brazil was entirely self-sufficient in the production of its pig iron needs and that by the early fifties self-sufficiency in ingot steel production had been achieved. This, however, would be an illusion. A detailed analysis of the imports of rolled steel products (which is not done here) would show that a large proportion of these products were of a semi-finished nature—products destined for rerolling operations in Brazil. The necessity for importing semifinished steel products would thus indicate that Brazil had not become entirely self-sufficient in pig iron or ingot steel production. The process of import substitution in rolled steel products, however, was much slower. Only with the full operation of Volta Redonda in the late forties did Brazil begin to rely on its own rolled products for nearly 70 percent of its needs. With the operation of Usiminas and Cosipa in the sixties, Brazil supplied itself with about 90 percent of the steel products it needed.

Besides observing the aggregate growth of Brazil's steel production, it is instructive to examine the changes that have taken place in the nature of Brazil's iron and steel production and consumption. In part a. of Table 21, we see that in 1965 more than 40 percent of Brazil's pig iron was still produced in charcoal-using furnaces and more than 3 percent in electric-reduction furnaces (the latter production took place in the two firms ACESITA and Mannesmann). Pig iron capacity based

Brazilian Pig Iron, Steel, and Rolled-Steel Production and Consumption, 1940-67

(tons)

	Pig Iron			Steel			Rolled-Steel		
	Pro-duction	Con-sumption	Import ÷ consumption	Pro-duction	Con-sumption	Import consumption	Pro-duction	Con-sumption	Import consumption
1940	185,570	185,999	—	141,201	147,810	4.5	135,293	414,519	69.4
1941	208,795	208,864	—	155,357	159,333	2.5	149,928	368,268	66.2
1942	213,811	213,837	—	160,139	161,743	1.0	155,063	262,764	43.3
1943	248,376	248,378	—	185,621	189,034	1.8	157,620	325,534	54.9
1944	292,169	292,730	—	221,188	259,350	14.7	166,534	492,613	68.1
1945	259,909	260,175	—	205,935	233,474	11.8	165,805	465,639	67.5
1946	370,722	371,837	—	342,613	378,824	9.6	230,229	656,751	65.4
1947	480,929	481,561	—	386,971	431,180	10.2	269,452	738,554	63.8
1948	551,813	551,815	—	483,085	492,545	1.9	381,480	567,579	38.2
1949	511,715	511,715	—	615,069	625,250	1.6	465,111	698,064	35.7
1950	728,979	—	—	788,557	803,119	1.8	572,489	843,049	32.6
1951	776,248	—	—	842,977	871,526	3.3	681,815	1,068,016	36.2
1952	811,544	—	—	893,300	911,831	2.0	703,103	1,087,934	35.6
1953	880,065	—	—	1,016,300	—a	—	794,460	1,006,821	21.1
1954	1,088,948	—	—	1,148,300	—	—	834,037	1,486,411	43.9
1955	1,068,513	—	—	1,162,500	—	—	932,283	1,265,659	27.3
1956	1,152,358	—	—	1,364,800	—	—	1,073,661	1,324,508	19.1
1957	1,251,657	—	—	1,470,000	—	—	1,130,189	1,521,321	25.9
1958	1,356,130	—	—	1,659,000	—	—	1,303,633	1,518,146	14.1
1959	1,479,742	—	—	1,866,000	—	—	1,492,009	1,998,826	25.3
1960	1,749,848	—	—	1,843,019	—	—	1,712,289	2,128,331	20.4
1961	1,976,230	—	—	2,443,221	—	—	1,931,785	2,257,701	14.8
1962	2,009,067	—	—	2,565,226	—	—	1,998,913	2,275,654	12.4
1963	2,374,963	—	—	2,824,045	—	—	2,142,000	2,631,700	18.6
1964	2,448,735	—	—	3,015,698	—	—	2,108,784	2,338,106	11.5
1965	2,340,637	—	—	2,982,994	—	—	2,096,815	2,305,860	8.6
1966	2,924,500	—	—	3,781,797	—	—	2,677,176	—	—
1967	3,057,084	—	—	3,696,145	—	—	2,853,177	—	—

Source: Same as Table 15; also *Boletim, IBS.*

a Consumption from that date on equal to domestic production.

Table 19

Indexes of Production and Consumption of Pig Iron, Steel Ingots, Rolled-Steel Production, and Industrial Production, 1940–66

(1949 = 100)

	Pig iron		Ingot steel		Rolled-steel products		Industrial Production
	Production	Consumption	Production	Consumption	Production	Consumption	
1940	36.3	36.3	23.0	23.6	29.1	59.4	52.6
1941	40.8	40.8	25.2	25.5	32.2	52.8	56.7
1942	41.8	41.8	26.0	25.9	33.3	37.6	58.2
1943	48.5	—	30.2	30.2	33.9	46.6	66.5
1944	57.1	—	36.0	41.5	35.8	70.6	70.1
1945	50.8	—	33.5	37.3	35.6	66.7	68.5
1946	72.4	—	55.7	60.6	49.5	94.1	78.3
1947	94.0	—	62.9	69.0	57.9	105.8	81.4
1948	107.8	—	78.5	78.8	82.0	81.3	90.6
1949	100.0	—	100.0	100.0	100.0	100.0	100.0
1950	142.5	—	128.2	128.4	123.1	120.8	111.4
1951	151.7	—	137.0	139.4	146.6	154.0	118.5
1952	158.6	—	145.2	145.8	151.2	155.8	124.4
1953	172.0	—	165.2	—	170.8	144.2	135.2
1954	212.8	—	186.7	—	179.3	212.9	146.7
1955	208.8	—	189.0	—	200.4	181.3	162.3
1956	225.2	—	221.9	—	230.8	189.7	173.5
1957	244.6	—	239.0	—	243.0	217.9	183.2
1958	265.0	—	269.7	—	280.3	217.5	213.2
1959	289.2	—	303.4	—	320.8	286.3	240.7
1960	342.0	—	299.6	—	368.1	304.9	264.8
1961	386.8	—	397.5	—	415.3	323.4	293.4
1962	393.7	—	417.5	—	429.8	326.0	316.0
1963	464.1	—	459.0	—	460.5	377.0	318.2
1964	477.9	—	490.0	—	511.1	371.8	334.1
1965	457.0	—	485.2	—	504.2	310.3	317.9
1966	597.0	—	615.0	—	620.6	—	—
1967	—	—	601.0	—	—	—	—

Source: Computed from data in Table 18 and from data in *Revista Brasileira de Economia*, Março 1966.

Table 20

Brazilian Production of Iron and Steel, 1916-66

(in tons)

Year	Pig iron			Steel ingot			Rolled-steel products		
	Production	Apparent consumption	Import/consumption (%)	Production	Apparent consumption	Import/consumption (%)	Production	Apparent consumption	Import/consumption (%)
1916	4,267								
1919	10,808								
1925	30,046	41,760	28.0	7,559	14,123	46.5	283	373,485	99.9
1930	35,305	37,258	5.2	20,985	24,766	15.3	25,895	259,224	90.0
1940	185,570	185,570	0.0	141,201	147,810	4.5	135,293	414,519	69.4
1945	259,909	259,909	0.0	205,935	233,474	11.8	165,805	465,639	67.5
1950	728,979	728,979	0.0	788,557	803,119	1.8	572,489	843,049	32.6
1960	1,749,848	1,749,848	0.0	1,843,019	1,843,019	0.0	1,712,289	2,128,331	20.4
1964	2,445,525	2,445,525	0.0	3,043,749	3,043,749	0.0	2,108,789	2,338,106	9.8
1965	2,258,529	2,258,529	0.0	2,978,122	2,978,122	0.0	2,096,815	2,308,860	9.1
1966	2,939,230	2,939,230	0.0	3,775,104	3,775,104	0.0	2,677,176		

Source: *A Economia Siderúrgica Da América Latina: Monografia do Brasil*, Comissão Econômica Para A América Latina, Santiago, Dezembro de 1964, mimeographed (monograph prepared by Dr. M. Falcão). *Boletim IBS*, Instituto Brasileiro de Siderúrgica. IBGE, *O Brasil em Números*, import data for iron and steel imports from Fundação Getúlio Vargas, IBRE.

on charcoal was even greater than production. The capacity figures, however, need an explanation. As is seen in Table 22, total output of pig iron in the mid-sixties varied between 50 and 65 percent of capacity of blast furnaces. This was because of substantial amounts of idle capacity in charcoal furnaces, especially the small furnaces of Minas Gerais. We thus have an explanation of the different distribution structure of production and capacity in pig iron.

The structure of steel production was still more than 50 percent based on the open-hearth (SM) method. The high proportion of electric steel furnaces is because of the large number semi-integrated steel plants, many of which produced special steels. The higher proportion of LD capacity than its proportional participation in actual steel output may be explained by the fact that Usiminas and Cosipa were advancing only slowly to their production capacity in 1965 (when these capacity measurements were made), while Mannesmann's LD furnace was working substantially below capacity.

It is interesting to compare the production levels and capacity

Table 21

a. Brazilian Pig Iron: 1965 Production and Capacity Distribution by Type of Installation

Type of installation	(Percentage distribution) Production	Capacity
Coke	55.4	51.8
Charcoal	41.2	45.4
Electric	3.4	2.8
	100.0	100.0

b. Brazilian Ingot Steel: 1965 Production and Capacity Distribution by Type of Installation

Type of installation	(Percentage distribution) Production	Capacity
Open Hearth (SM)	56.2	43.9
L. D. (Oxygen)	26.0	34.5
Electric	17.8	21.6
	100.0	100.0

Source: Computed from estimates made by Tecnometal and Booz Allen & Hamilton International for the BNDE.

Table 22

Capacity and Output of The Brazilian Iron and Steel Industry

(in 1000 tons)

I—*Totals for Brazil*

a. *Pig iron*

		Output	Capacity	Output as percentage of 1965 Capacity
	1964	2,446		54
	1965	2,259	4,541	50
	1966	2,939		65

b. *Steel ingot*

	1964	3,044		60
	1965	2,978	5,078	59
	1966	3,775		74

c. *Flat products*

	1964	902		15
	1965	1,019	5,927	17
	1966	1,359		23

d. *Nonflat products*

	1964	1,207		43
	1965	1,078	2,797	38
	1966	1,317		47

II—*Individual firms*

a. *Pig iron*

	Output			Capacity	Output as a percentage of 1965 Capacity		
	1964	1965	1966	1965	1964	1965	1966
Volta Redonda (CSN)	957	927	875	1,020	94	91	86
Cosipa		43	401	565		8	71
Usiminas	276	382	505	575	48	66	88
Belgo-Mineira	390	338	422	440	72	63	78
Mannesmann	170	122	80	280	57	44	29
Acesita	63	76	83	125	38	46	50

b. *Steel ingot*

	1964	1965	1966	1965	1964	1965	1966
Volta Redonda (CSN)	1,218	1,256	1,248	1,400	87	90	89
Cosipa		30	431	625		5	69
Usiminas	276	383	529	634	43	60	83
Belgo-Mineira	421	409	475	450	93	91	105
Mannesmann	214	195	198	328	65	59	60
Acesita	82	91	103	120	68	76	86

Table 22 (Continued)

c. *Plates*

	Output—1965	Capacity—1965
Volta Redonda	97	638
Cosipa	21	800
Usiminas	140	800
d. *Hot rolled sheets*		
Volta Redonda	176	525
Cosipa	66	600
Usiminas	23	750
Acesita	27	43
Belgo-Mineira	61	105

Source: Boletim IBS, Instituto Brasileiro de Siderurgia; *Siderurgia-Metalurgia-Mineração*, Editora Banas, S.A., 1967; unpublished reports of Booz Allen & Hamilton to the BNDE and special study of firm Tecnometal for BNDE.

estimates reproduced in Table 22. It should be made clear that when building an integrated steel mill for flat products, the production capacity of the rolling mills usually far exceeds the capacity of the blast furnace and steel shop sections. This is because of the indivisible nature of most rolling-mill installations. The blast furnace and steel shop sections, by their very nature, can be built up on a much more gradual basis.

Before commenting on the results presented in Table 22, a brief clarification should be given of the capacity measurements of the rolling mills. The huge amount of excess capacity in the rolling mills is, as explained above, partially because of the lumpiness of investment in them, and only with further investment in blast furnaces and steel shops, which grow more in accordance with demand, will they be more fully used. The degree of excess capacity is also slightly exaggerated however. Suppose that a steel mill has one rolling mill with capacity of 200 tons and one with 100 (rolling mills producing different types of products). This does not necessarily mean that total effective rolling mill capacity is equal to 300 tons. The nature of the market might be such that there is perfect demand complementarity between the products of the two rolling mills. Thus, assuming no other firms producing similar products and no possibility for importing, total effective capacity for rolled steel products is only 200 tons. The extra 100-ton capacity of the first rolling mill will only become effective capacity when new investments will have been made to enlarge the second rolling mill's capacity to 200. With these considerations, the effective capacity of Brazil's rolling mills is not quite as large as it appears on Table 22.

The capacity figure for the blast furnaces should also be taken with some qualifications. Blast furnace capacity is extremely flexible; through the use of beneficiated ores and/or injection of fuel oil, many blast furnaces have wound up producing substantially more than they originally were expected to produce.

Table 22 shows that in the blast furnace sections and steel shop sections, Brazil worked closer to capacity than in the rolling mills. Much of the excess capacity in blast furnaces for Brazil as a whole is a result of small independent blast furnace operations which flourished in the fifties when there was a general shortage of pig iron (see previous chapter), but many of which are currently standing idle.

THE POSTWAR RATE OF GROWTH OF STEEL PRODUCTION AND CONSUMPTION AND THE GROWTH OF THE BRAZILIAN ECONOMY

A comparison of the growth rates of ingot and rolled-steel production with various aggregates is shown in Table 23. It will be noted that the yearly growth rate of the real gross domestic product in the period 1947–61 was smaller than the yearly growth rate of industrial production, and that the yearly growth rates of steel-ingot production and rolled-steel-products output were substantially higher than the growth rates of industrial production. Thus, iron and steel production was one of the leading subsectors of industry, the leading sector of the economy. These growth rates cannot be dismissed as being high because of a tiny base. Table 18 shows that iron and steel production at the beginning of the post–World War II period was not negligible. Also shown on Table 23 are the growth rates of apparent consumption of steel-ingot and rolled-steel products. These are smaller than the production growth rates, which obviously reflects a substantial amount of import substitution in steel products. As recently as 1961–64, the apparent consumption growth rate of ingot steel was still larger than the industrial production growth rate, while the apparent consumption growth rate of rolled steel products was just slightly smaller than the latter. The industrial production index includes, however, a substantial proportion of production directly substituting imports. Taking this into account, the rate of growth of steel consumption, which represents a growth rate without direct import substitution, may still be taken as an indicator that the steel industry is one of the leading net growth industries within the general industrial sector.

The high growth rates of steel consumption in Brazil could be attributed to the policies adopted by the government to maximize the vertical integration of the industrialization process within the country. There exists substantial evidence to show that when most linkages of an industrialization process occur within a country (this is another way of describing vertical integration), the impact on steel production is very strong relative to other sectors affected. It has been shown, for instance, that when ranking the degree of forward linkage of fourteen major industrial sectors in the United States, the iron and steel industry came out first, and when ranking backward linkages, the iron and steel industry came out fifth.[1] In absolute terms, the power of forward linkage was about twice as strong in iron and steel than in the industrial group which ranked second, while in the backward linkage measurement (where iron and steel ranked fifth), the industry was less than 15 percent behind the first-ranking industry. The total repercussion, a weighted average of forward and backward linkages, gave iron and steel the highest ranking. This particular exercise was done for the United States, which is a fairly self-sufficient economy in which domestic linkages are very strong. Brazil's effort at producing a fairly self-sufficient industrial complex, maximizing internal linkages, is thus an attempt to emulate the United States experience. The relative *ranking* of industries in terms of linkage effects may thus be considered similar to both economies (of course, not the absolute values, which are not comparable).

Another insight into the high growth rates of steel relative to other sectors of the economy may be gained from the well-known United Nations study on industrial growth.[2] Taking a combined cross-section of fifty-three countries in 1953 and forty-eight countries in 1958, a regression analysis was made to determine the influence of per capita income and population on the degree of industrialization; this was followed by a regression analysis to determine the value of output of thirteen different sectors, using per capita income, population, and degree of industrialization as independent variables. It was found that the income elasticity of total industrial output was 1.37, while the income elasticity for the basic metal sector was 1.99 (this number was

1. Baer, *op. cit.*, pp. 138–144.
2. United Nations, *A Study of Industrial Growth* (New York, 1963).

surpassed only by the paper-products sector which reached 2.03). Basic metals had the highest population elasticity and also the highest proportional response to the variable "degree of industrialization." The very high response of the basic metal sector in the cross-section study makes the high growth rate of the Brazilian steel sector in relation to industrial production and GDP growth rates consistent with observed trends throughout the world.

As a matter of fact, an experiment made for the year 1962, when Brazilian data for per capita income, population, and degree of industrialization were used to "predict" the "normal" pattern of Brazil's industrial structure within the framework of the U.N. model, showed that the actual contribution of the basic-metals sector was much smaller than the expected one.[3] It was predicted that basic metals would make a contribution of 9.7 percent to the total value added of the industrial sector, while the actual contribution only amounted to 5.7 percent. This divergence could be because the basic metals sector is not confined to the iron and steel industry.[4] It should also be taken into account, however, that in 1962 the import of steel products amounted to about 15 percent of total steel products consumption as measured in tonnage. The actual and normal patterns were measured in terms of value added, and it should be noted that the 15 percent of steel products imported in that year were of a specialized nature, with a much higher value per ton than the average produced in Brazil.

Whatever the value of this exercise in terms of accuracy, the results indicate that Brazil's growth of steel output was not out of proportion to general industrial growth and that one might even detect the possibility that the sector should have grown at an even faster rate.

Further evidence of the possibility of the continued growth of steel production in Brazil may be obtained by an examination of Table 24,

3. Baer, *op. cit.*, p. 145.

4. Another study, which used the same UN model, calculated actual and predicted values for the years 1949 and 1964. For 1949 it was found that the actual value for Basic Metals products was 6 percent and the predicted 5 percent; and for 1964 the actual value was 9 percent and the predicted one 10 percent. See Joel Bergsman and Arthur Candal, "Industrialization: Past Success and Future Problems," in *Essays on the Economy of Brazil,* edited by Howard S. Ellis. The higher actual value here for 1949 could be explained by the fact that only two years before Volta Redona started to function, while the industrial spurt of the fifties had not yet started. The differences found in 1962 and in 1964 could be explained by the fact that in 1964 such newer firms as Usiminas and Ferro e Aço de Vitoria had come into operation, thus bringing the actual participation of the basic metals sector more closely into line with the predicted participation.

Table 23

Yearly Growth Rates

	Real GDP	Industrial production	Steel-ingot production	Rolled-products production	Steel-ingot apparent consumption	Rolled products apparent consumption
1947–61	6.00	9.50	13.50	15.30	12.50	9.25
1947–55	5.75	9.00	14.50	16.00	13.50	8.50
1956–61	6.75	11.00	10.50	12.50	11.00	11.00
1961–64	3.50	4.40	10.00	7.25	10.00	4.80

Source: Calculated from same sources as mentioned in Table 18.

where I have made a comparison of per capita ingot steel consumption
for selected Latin American countries. It will be seen that despite the
fact that Brazil is Latin America's major steel producer, ingot steel
consumption per capita is substantially inferior to Argentina, Chile,
and Mexico, and even smaller than the average for Latin America.
(United States per capita steel consumption is about 615 kgs.) It can be
inferred from this that continued industrialization and per capita income growth in Brazil should lead to substantial increases in the market
for steel and thus in the expansion of steel production.

Besides observing the aggregative growth rate of steel production, it
is instructive to observe the changes that have taken place in the
nature of Brazil's steel production and consumption. It was already
mentioned above that steel products may be divided broadly into flat
and nonflat products, the latter predominating in construction activities
and the former in manufacturing. One would thus expect an industrial-
izing country to use an increasingly higher proportion of flat products.
In Table 25, which shows the proportion of flat products produced and

Table 24

Per Capita Apparent Steel Consumption and Gross Domestic Product for
Selected Countries

	(in kg per capita and US$ per capita)					
	Apparent per capita steel ingot consumption			Gross domestic product per capita		
	1952	1964		1953	1964	
Brazil	26.2	44.2	(3,222)ᵃ	250	325	
Argentina	46.7	108.0	(1,833)	—	616	
Colombia	13.0	33.5	(264)	—	301	(1958)
Chile	47.5	86.4	(611)	—	411	(1958)
Mexico	34.1	61.6	(2,402)	232	400	
Peru	14.4	23.7	(87)		250	
Latin America (total)	29.3	49.5	(8,793)			
USA		615.0	(118,067)	2,080	2,900	

Sources: United Nations, *Statistical Yearbook*, 1965; United Nations, *Yearbook of National
Accounts Statistics*, 1964; Comision Economica para America Latina, "La Economia
Siderurgica de America Latina," February 1966, mimeographed.

a. Figures in parenthesis represent total ingot steel production in thousands of tons;
the Brazilian figures are not accurate, but were not changed since the CEPAL document
did not mention how the estimates were arrived at for each country; these aggregates
should, however, give an idea of the relative magnitude of each country's steel capacity.

consumed as related to total production and consumption, we find the
expected trend. Until the late forties, hardly any flat products were
produced in Brazil, but by the sixties more than 40 percent of total
output in Brazil consisted of flat products. The proportion of flat products consumed also changed drastically after World War II. Their
rise is easily explained by the nature of Brazil's industrialization in
the fifties and sixties. Since the emphasis was on the stimulation of
such flat-product-using industries as automobiles and shipbuilding and
on the vertical integration of economic activity, one should have expected these changes in the structure of steel production and consumption.[5]

Table 25

Production and Consumption of Flat
Products as a Proportion of Total
Production and Consumption of
Rolled-Steel Products

	Production	Consumption
1925	0.0	22.8
1930	0.0	25.0
1940	0.0	35.3
1945	0.0	26.5
1950	34.4	34.5
1960	41.3	41.6
1964	40.8	41.2
1965	48.0	—

Source: Calculated from same sources as
Table 15.

Table 26 contains estimates of the sectoral distribution of steel
sales. Part a. presents a distribution of steel product sales in Brazil in
1956 and in 1960, and of steel product sales distribution in the U.S. and
Europe at various periods of time. The estimates for Brazil in 1964
and 1965 in parts b. and c. of the table are not exactly comparable, since
the classification and breakdowns used were different. A comparison of

_____

5. Steel experts have told me that the long-run equilibrium ratio of flats to
nonflats for an industrialized country is about 50:50. As industrialization proceeds,
at first the use of flats rises much faster than that of nonflats. But once a certain
level of industrialization is reached, a country tends to consume more and more
nonflat special steels. Acording to the calculations of Hans Mueller, the production
of flats amounted to 48.9 percent in the European coal and steel community in 1965,
and in the United States the proportion was 60 percent in 1960.

all three parts of the table give an idea of the trends, however. All the estimates for the automobile sector's share, except for 1965, show that it was still substantially behind the American and European shares of that sector. The Booz Allen estimates for 1965 seem rather high in comparison with the automobile estimates for other years, which leads me to assume that they included some of the indirect steel consumption of that industry. Construction in the late fifties absorbed a higher proportion of steel products than the more advanced countries. This would seem natural in a country undergoing rapid industrialization and urbanization. The decline of the proportion of steel going to that sector, as shown by the 1965 estimates, however, should be related more to the stagnation of investment and construction in the mid-sixties rather than to long-term trends towards a more "normal" American-European type of structure. The low absorption rate of the machinery sector compared with the other industrial countries reflects the smaller capital goods sector of Brazil's industrial complex compared with that of the Uinted States and Europe.

The main consumers of flat products are the automobile, naval, petroleum, packing (tin cans), and a large proportion of the metals products industries; of course, construction is one of the principal customers for nonflat products, but the "permanent way" section of the railroad industry (production of rails) and the machinery industry also absorb a substantial amount of nonflat products. If we assume that the industrial growth of Brazil into the seventies will produce an industrial structure similar to that of the United States in the mid-fifties, one should expect a substantial spurt of demand for flat products from the automobile, canning, and container industries. An additional source of demand for flat products should also come from the shipbuilding, where in relative terms Brazil might approximate more some European countries than the United States. Construction will be the most important future source of demand for nonflat products, while the machinery sector is also bound to increase as an absorber of nonflats, if import substitution in the capital goods sector will continue grow.

Looking at part c of Table 26, which contains a finer breakdown of sectors, it would seem that the weight of the highway-equipment and agricultural-equipment sectors will be more pronounced in future demand for steel products than was the case in the mid-fifties, while the weight of railroads should decrease. Since the development of much of the Brazil's interior is being accomplished by road, the demand for

Table 26

Sectoral Distribution of Steel Products

(percentage distribution)

	Brazil		United States of America		United Kingdom		West Germany	Western Europe
	1956	1960	1928	1956	1936	1955	1955	1956
a. International Comparisons								
Machinery and Equipment	7.3	8.4	16.2	14.8	26.4	15.1	18.9	17.0
Transport	8.3	9.3	24.1	26.1	17.7	24.2	19.0	20.0
a. naval	—	0.6	0.9	1.3	6.5	7.7	5.8	5.0
b. rail equipment	4.0	1.6	4.3	4.1	3.3	5.3	1.9	4.0
c. automobile	4.3	7.1	18.9	20.7	7.9	11.2	11.3	11.0
Metal Products	40.0	34.5	23.7	29.8	26.6	29.8	27.9	22.0
(domestic appliances)	(4.7)	(3.1)	(2.4)	(5.0)	—	—	(3.1)	(9.0)
Construction	32.0	33.9	26.2	27.1	22.9	28.0	30.6	28.0
Railroads	7.6	7.9	9.8	2.2	6.4	2.9	3.6	3.0
Other	4.8	6.0	—	—	—	—	—	—
Total	100.0	100.0	100.0	100.0	100.0	100.0	100.0	100.0

Source: EPEA, Ministério do Planejamento e Coordenação Econômica, *Siderurgia, Metais Não Ferrosos, Diagnostic preliminary,* Rio de Janeiro, Abril 1966.

b. *Estimates for 1964*

Automotive	6.4
Construction	27.9
Railroads	4.4
Naval	6.9
Machinery	6.3
Metallurgical Ind.	31.5
Other	16.6
Total	100.0

Source: Estimated from data of Tecnometal study for BNDE (unpublished).

c. *Booz Allen Hamilton 1965 Estimates*

Automotive	12.6
Railroads	
a. rolling stock	.7
b. permanent way	6.7
Shipbuilding	2.0
Highway equipment	.6
Agricultural equipment	1.3
Canning	8.7
Containers	3.9
Domestic appliances	3.6
Commercial equipment	1.3
Construction	26.1
Industrial machinery	6.9
Job shops	8.1
Drawing mill	13.8
Miscellaneous	3.7
Total	100.0

Source: Unpublished Booz Allen & Hamilton International report to World Bank and BNDE, see bibliography.

highway equipment should continue to rise, and a larger proportion can and probably will be produced in Brazil. Also, assuming substantial modernization in agriculture in the final years of the sixties and throughout the seventies, one may expect a greater production of agricultural equipment than in the past. It seems obvious that the role of railroads in the future expansion of Brazil's transport network will be smaller than in the past. Taking this trend into account, one would expect the role of this sector to decrease within the steel market of the country.

The general conclusion one comes to when looking at the sectoral distribution of steel products is that the main stimulant to the continued growth of the steel industry within the country is a rapid rate of growth of the industrial and construction sectors.

It is not surprising that the steel industry should have a most powerful forward linkage effect, since its output is rarely a final product but an input into another industrial sector. If one considers that until the mid-sixties most of Brazil's steel output was used domestically, the full linkage effect of the industry was felt internally. It is probable that the repercussion or linkage effect on the economy of flat products is more pronounced than that of nonflat products, since these products will undergo more transformations in the process of being turned into final products and thus add more value than such items as rails or construction rods. Thus, the larger the value of flat products produced, the greater will be the repercussion effect of steel-production on the economy.

EMPLOYMENT EFFECT OF STEEL

The total direct employment effect of the iron and steel industry in 1964 was estimated at 80,000 workers,[6] which represented about 4 percent of the industrial labor force of Brazil and approximately 33 percent of the work force engaged in the metallurgical industry as a whole.[7] This is an impressive number when one considers that the

6. Estimates based on Tecnometal study and on direct information received from firms.

7. If we want to broaden the concept of the direct employment effect of the iron and steel industry by including the employment created by the main direct consumers of steel products, the total employment created amounts to between 550,000 and 600,000 workers, which represents close to 30 percent of Brazil's industrial labor force. These estimates are derived from information contained in *Inquéritos*

United States which produced more than forty-two times as much ingot steel as Brazil in 1964, employed only seven times as much labor.

A series of qualifications, however, are in order. This employment estimate includes many of the very small iron and steel firms, wherein production methods are very labor intensive but which contribute only a small proportion of total output. About 55 to 60 percent of Brazil's iron and steel workers were listed as being engaged in auxiliary services and administration and only the balance worked directly on the production line.

An examination of the employment structure of various steel firms showed that the proportion of workers engaged directly on the production line varies substantially from firm to firm. For instance, in 1964 only 28 percent of the workers of Usiminas were engaged directly on the production line, while more than 23 percent were working in "administration." This represented an oversupply of bureaucratic workers, since the average for administration in Brazilian steel firms was about 14 percent, and some smaller firms brought this proportion down to less than 5 percent.[8] There are three explanations for the high proportion of auxiliary workers in such firms as Usiminas and Volta Redonda. First, as was already mentioned in Chapter Two, the workshop sections of Brazilian steel firms are larger than in more advanced industrial countries because of the spare-parts-supply problem. A workshop is usually very labor intensive in nature. Second, the foundry section producing ingot molds also absorbs more labor in Brazil than in more automatized foundry shops in the United States. Third, in the case of Usiminas and Cosipa, a substantial amount of construction was still taking place in 1964 and 1965, which swelled the ranks of workers in the "auxiliary sections."

Of course, there can be no doubt that overemployment did exist in office staffs and all sorts of marginal services. It is interesting to note, however, that a certain degree of featherbedding was not restricted to government firms. Such private firms as Belgo-Mineira and Mannessmann had only 39 and 43 percent of its respective labor forces working directly on the production lines.

Econômicos—1966, Rio de Janeiro, IBGE—Conselho Nacional de Estatistica, Grupo Especial de Trabalho Para as Estatísticas Industriais, Junho 1967.

8. Usiminas became aware of its overstaffing problem and in 1965 hired the firm Booz Allen & Hamilton International to make a study to streamline the firm.

It is clear from an examination of different types of firms that the larger ones have usually a larger proportion of their workers engaged in auxiliary services. Two explanations may be offered for this phenomenon. First, the big integrated firms need more services than smaller and less complex operations. Second, the larger firms are often far from settled urban centers (with the exception of Cosipa) and have to employ a considerable labor force to run a large number of social services. Furthermore, Brazilian social legislation forces large firms to provide a whole array of social services for its workers.

Table 27 offers a comparison of steel ingot output per worker between the United States, Brazil, and Volta Redonda. I am not defining this measure as labor productivity because, as I will argue in the following chapter, the concept of labor productivity in the steel industry has little meaning. It will be noted that the output of steel ingot per person employed in the iron and steel industry is from five to six times higher in the United States than in Brazil. This contrast has to be qualified somewhat by taking into account that a larger proportion of the total iron and steel labor force in Brazil works in smaller enterprises than in the United States (such as pig iron works which do not produce steel, and iron and/or steel foundries). It is therefore instructive to look at the ingot output per worker in Volta Redonda, which has been established longer than the other larger firms and which is slightly

Table 27

Ingot Steel Output Per Man Employed

	United States	Brazil (total)	Volta Redonda
	(tons)		
1949			36
1950	163		48
1955	187		64
1960	174		87
1963			101
1964	229	38	93[a]
1965			95[a]

Sources: Same as Table 14; also calculated from *Boletim IBS*; Cotrim, N.C.B., *A Indústria Siderúrgica No Brasil e a Necessidade de Formação de Pessoal Técnico para Atender a Sua Expansão,* Volta Redonda, 1966; U. S. numbers calculated from the *Annual Statistical Report* of the American Iron and Steel Institute.

a. The decline in 1964 and 1965 is due to the general economic recession which took place in those years, lowering the production level of the steel industry, without lowering the work force of Volta Redonda.

more comparable to the average American steel enterprise. Here there has been a substantial growth of ingot steel per employed worker, reaching close to 45 percent of the American figure in the mid-sixties.

The reason for the low initial figures for Volta Redonda is that when the plant was finished in the late forties, a large proportion of the construction work force was kept on to be used in various sections of the steel enterprise. It is true that there were sociopolitical pressures for this. To a certain extent, however, a large proportion of the work force in a steel mill can also be considered more nearly a fixed than a variable input. A large steel establishment needs a certain minimum number of workers around if all sections are to operate, although they may operate substantially below full capacity. As output grows, the work force does not have to grow proportionately, thus raising production per man employed. One authority has estimated that, given past trends, the output of ingot steel per man employed should rise by the mid-1970s to about 140 for Volta Redonda.[9]

LABOR SUPPLY: ENGINEERS

As mentioned in the previous chapter, the need for engineers for the iron and steel industry was anticipated many years ago by the creation of a number of engineering schools, especially in São Paulo, Minas Gerais, and the army's special school. This does not mean, however, that the supply was abundant. In the United States the norm is a ratio of 1 engineer for every 50 workers. At Volta Redonda in the mid-sixties the ratio was 1 engineer for every 63 workers, and for the Brazilian steel industry as a whole the ratio was 1 for every 112. For the United States, the normal relation of technician to worker in the steel industry is 1 to 12.5; at Volta Redonda it is 1 to 42; and for Brazil as a whole it is 1 to 75.[10] Of course, interpretation of these data should be somewhat qualified because in many parts of the Brazilian firm (even Volta Redonda), especially outside the main production lines, there is a greater use of labor than in the United States. This labor need not, however, necessarily work under the close supervision of engineers or other highly trained personnel.

9. Cotrim, N.C.B., *A Indústria Siderúrgica No Brasil e a Necessidade de Formação de Pessoal Técnico para Atender a Sua Expansão*, Pamphlet published by the Companhia Siderurgica Nacional, 1966, p. 25.
    10. *Ibid.*, p. 12.

At Volta Redonda, about 1.5 percent of the total labor force consists of engineers, and almost half of these are metallurgical engineers and the rest are in the fields of electrical, mechanical, and civil engineering.[11] If we assume that the proportion found at Volta Redonda should be the norm for Brazil, the steel industry in the mid-sixties should be employing approximately 1,200 engineers, half of which would be metallurgical engineers. If we assume that by the mid-1970s the industry will more than double its output, but only double its labor force to 160,000, the number of engineers needed if the proportions were not to change would be 2,400 (again, half of which is in metallurgy). Thus, in the decade from the mid-sixties to the mid-seventies, about 120 new engineers would have to be graduated every year just to satisfy the needs of the steel industry, and about half of these would have to be metallurgical engineers. Since, however, in the mid-sixties the Volta Redonda proportions of engineers to total workers were higher than those of the remainder of the steel industry as a whole, the yearly demand for engineers to attain the 1.5 percent ratio in the mid-seventies would have to be substantially greater than 120.

Table 28 gives an idea of Brazil's manpower training in engineering. In 1963, out of a total of 489 non–civil engineers graduated, a few more than a hundred were graduated in metallurgical engineering. Assuming that the necessary absorption of engineers until the middle of the seventies will be 120 to 150 per year in the steel industry and half of these will be metallurgical engineers, the training in terms of aggregate output would seem to be adequate. Since the iron and steel industry only employs one third of the work force in the entire metallurgical sector, however, it is doubtful that the steel industry will be able to claim some two thirds of all the metallurgical engineers graduated. Although the growth of engineering training from the early forties has been very high, it will be noted in Table 28 that total enrollment in engineering in Brazil is still extremely low compared with most industrialized countries. The rise of the proportion of non–civil engineers trained, however,

11. In 1965, Volta Redonda's engineers were divided among the following special fields:

Metallurgical	48.96%
Mechanical Electrical	24.90
Civil	15.77
Others	10.37
	100.00%

Table 28

Training of Engineers in Brazil

a. *Enrollment in Engineering Schools*[a]

	Civil	Other	Total
1954	5,357	1,839	7,196
1960	5,342	4,070	9,412
1963	4,357	3,622	7,979

b. *Enrollment in Engineering Schools in Other Countries in 1959*

United States	278,330
UK	16,910
Sweden	4,770
W. Germany	31,100
France	15,000
Turkey	16,910
Yugoslavia	19,620
Greece	1,700

c. *Graduating Engineers in Brazil*

	Civil	Other	Total
1940	178	75	253
1950	696	171	867
1960	731	546	1,277
1963	713	489	1,202

Sources: Educação, II, EPEA, Ministério do Planejamento e Coordenação Econômica, Plano Decenal de Desenvolvimento Econômico e Social, Rio de Janeiro, 1966; international comparisons taken from Leff, Nathaniel H., *The Brazilian Capital Goods Industry: A Case in Industrial Development* (Harvard University Press, 1968).

a. Does not include Chemical engineering.

reflects the increasing use of metallurgical, electric, and other specialties with the increase of Brazil's industrial complex.

Although the supply of engineers has not been adequate, the industry has never experienced a deficit of bottleneck proportions. A substantial amount of adaptation and improvisation took place. Some jobs reserved in more advanced countries for engineers were often taken over by highly skilled technicians in Brazil. In a number of firms, civil engineers took over functions which would normally be undertaken by metallurgical, mechanical or electrical engineers.[12]

Most of the major steel firms have special training programs for their engineers, both within the company and at universities. Volta

12. Cotrim, *op. cit.*, p. 16.

Redonda even established its own engineering school. Brazilian engineers have proved very adaptable. Most of the major firms were started with a large component of foreign engineers supervising the major sections. Within a relatively short period, however, Brazilian engineers took over most of the tasks. Many Usiminas engineers, for instance, were sent to Japan for specialized on-the-job training, while many Cosipa engineers were sent to the United States and Europe, because of the special connections the firm had with the Kaiser Company. Of course, many Mannessmann engineers received their special training in Germany.

There can be no doubt that such latecomers as Usiminas and Cosipa benefitted by the pioneering efforts of Belgo-Mineira and Volta Redonda. These firms built up a pool of experienced steel engineers, some of whom were bid away by the newer firms, but some also were lent out in order to give assistance.

LABOR SUPPLY: SKILLED AND UNSKILLED WORKERS

Out of Volta Redonda's total work force, a little more than 17 percent were unskilled, about 21 percent semiskilled and a little more than 35 percent skilled workers. Industrywide use of skilled labor was lower than that of Volta Redonda. If we assume, however, the Volta Redonda proportion for the entire iron and steel industry, approximately 28,000 steel workers in the mid-sixties would be considered skilled. Assuming the doubling of the work force by the mid-seventies and the same proportions, 28,000 additional skilled workers would be needed by the steel industry, or an annual addition of 2,800 new skilled workers. In 1963 and 1964 respectively, 5,362 and 6,241 students were graduated from secondary industrial schools, of which 2,847 and 3,382 were in the states of São Paulo, Minas Gerais, and Rio de Janeiro. In 1965, about 62,000 students were enrolled in industrial schools; these were distributed over an eight-year schooling range. Given the fact that in the mid-sixties a substantial deficiency already existed in the number of skilled workers available to the steel industry, the shortage of such labor as evidenced by the gap in supply and demand is quite substantial.

The deficiency in Brazil's training system has forced firms to improvise—to place semi- or unskilled labor into positions substantially above what would be commensurate with their training. Of course, this implies a substantial amount of on-the-job training. Most firms have also instituted special courses for workers already employed.

The impression I had from touring Brazilian steel plants and from talking to engineers and supervisors in various sections of steel mills is that Brazil's unskilled labor has adapted very rapidly to the tasks involved in various parts of a steel mill. Most firms conduct special tests for new workers, and those who are naturally adept at handling machines, and/or who have quick reactions, are, for example, rapidly trained to handle the switches that control the passing of ingots through the strip mill.

In one of the firms I visited, a foreign supervisor told me that after a few months there was scarcely any difference between the workers he directed at blast furnaces and steel shops in Europe and the workers doing the same tasks in Brazil. He pointed to two principal defects, however, which may be expected in a first-generation work force. First, he noted that workers had "less respect" for machines in Brazil than in Europe, resulting in more breakage. Second, the Brazilian worker was not in the habit of "thinking ahead." For instance, a worker, and often a supervisor, would not anticipate the need to refurbish the diminishing stockpile of some essential input but would wait until it was all gone before calling attention to the need. This would often prejudice the operation of an entire section of the steel mill. Of course, at times this trouble could also be blamed on faulty production planning.

All the large steel firms have extensive training programs, extending from primary schools to the end of secondary schools (and, in the case of Volta Redonda, including an engineering school). Most firms also conduct adult-education programs. The training programs are legally required of most larger firms and are run in co-operation with the SENAI (National Industrial Apprenticeship Service).[13] This service was founded in the early forties in an effort to cope with the severe shortage of skilled labor in Brazil. To finance the SENAI, the government instituted a one percent payroll tax.

For a number of reasons, the SENAI system has not developed fast enough to provide enough skilled workers to meet the demand of Brazil's growing industrial complex.[14] The tremendous expansion of the country's steel-producing capacity in the fifties and early sixties has made the supply of skilled steel workers especially scarce and forced most firms to place semiskilled workers into skilled positions, relying mainly on

13. For a more thorough description of the SENAI system in Brazil see Nathaniel H. Leff, *The Brazilian Capital Goods Industry, 1929–1964,* Chapter Three.
14. *Ibid.*

an on-the-job training system. The creation of training programs in each new steel firm, however, should in the future make up for the initial difficulties. Of course, many of the firms located in more isolated parts of Brazil face the danger of providing industrial training to young workers who will then migrate to urban centers where they can get more highly skilled jobs and enjoy the more varied life of larger cities.

Even where a decent training system has been established and where there are no excessive shortages of skilled workers, the type of training received and the use of such workers often leaves a lot to be desired. For example, one authority has claimed that many Brazilian steel firms do not appreciate the importance of the foreman and supervisor in influencing the performance of workers on the production line.[15] Promotion to foreman is often simply based on performance on the job, and little effort is made to give them a more general training which would allow them to adapt more rapidly to new methods of production and transmit these effectively to the men they supervise.

The general picture that emerges is that the iron and steel industry of Brazil has had a substantial employment effect on the economy. A sufficient number of trained engineers has always been on hand to take over the direction of steel operations in a fairly short period. The availability of skilled manpower was very small, but the assimilation of workers, the degree of on-the-job training which took place, was relatively rapid. It would be fair to say that the manpower training which the growth of the industry represented to the country (the learning-by-doing and the official training given by the company-SENAI schools) probably more than outweighed the loss of efficiency caused by the large proportion of inadequately trained and inexperienced labor force the steel firms had to employ. Of course, to measure this would involve a separate study beyond the scope of the present work. It will be seen in the following chapter, however, that in the case of the steel industry, because of the technology of steel production and the relatively small labor input, the concept of labor productivity is not very meaningful in judging the performance of the industry.

DEVELOPMENT OF THE STEEL EQUIPMENT INDUSTRY

Another important backward linkage effect of the growth of Brazil's

15. Cotrim, *op. cit.,* pp. 18–19.

iron and steel production has been the development of the steel equipment industry. It was noted in Chapter Two that by the mid-sixties the import content of new investment in the steel industry was substantially lower in the expansion plans of the decade reaching into the mid-seventies than the previous decades. This reflects a substantial expansion of firms producing equipment for the steel industry, such as cranes, cylinders for rolling mills, compressors, hydraulic presses, parts for the construction of blast furnaces, and electric furnaces. It is true that these firms do not necessarily specialize in equipment for the steel industry and that many existed already before the fast post–World War II growth of the industry. Their substantial expansion, however, was strongly influenced by the needs of the rapidly expanding steel industry.

By the mid-sixties, Brazilian steel foundries could produce pieces of equipment up to 30 tons and the iron foundries up to 60 tons. Installations for the production of the cylinders for large rolling mills (weighing up to 50 tons), for example, were still not available, however. The equipment industry in the mid-sixties consisted of about twenty-two firms, employing nearly 34,000 workers. This high employment effect resulted from the highly labor-intensive nature of the work involved. It is thus obvious that in the future the expansion of the steel industry will have a larger internal impact than in the past. Of course, a number of important firms in the equipment industry are foreign (such as Ishikawajima do Brasil, Indústria Elétrica Brown-Boveri, S.A.), and others work under foreign licenses. Thus, an expansion of the steel industry, which would rely more than before on domestically produced capital equipment, would still create a foreign exchange leak through the profit remittances of foreign firms in the equipment business and the payment of royalties by firms operating under foreign license.[16]

16. Comments in this section and numerical estimates are based on materials of a special study by the firm Tecnometal for the BNDE, entitled "A Industria de Produção de Equipamentos para a Siderúrgia."

6

PERFORMANCE OF BRAZIL'S
STEEL INDUSTRY: INTERNAL

MY ANALYSIS of the performance of Brazil's steel industry will fall into two parts. In this chapter, I shall consider what might be called the "internal performance" of the industry, an analysis of the industry's productivity in iron and steel, estimates of the cost of production, and an international comparison of steel costs and prices. In the following chapter I shall examine the industry's "external performance"—the advantages and disadvantages of the industry's locational structure and the "opportunity cost" which the development of the industry implied for the economy.

PRODUCTIVITY

A close examination of the steel industry has led me to the conclusion that an analysis of productivity of labor is not very relevant in trying to evaluate the performance of the industry. I have found in my cost estimates that direct labor input in the blast furnaces of large integrated mills never amounted to more than 2 to 3 percent of the total cost of one ton of pig iron. In smaller establishments, direct labor costs were about 5 to 7 percent of the total cost of pig iron. This proportion was not larger for steel furnaces, and even smaller, fluctuating around 1 percent, in the rolling mills.[1] This implies that even large

1. Another economist, studying labor productivity in the steel industries of various European countries, came to the same conclusion. He found that "in the steel industry labor cost is only a rather small part of total cost" and that "the figures on labor requirements are certainly elements of total efficiencies by which the steel industries could be compared, but they do not by themselves give the complete

proportional variations in the labor input price (ranging from 50 to 100 percent) will not have an appreciable effect on costs. I am not trying to say that total labor employed does not have a substantial bearing on total costs of the firm. There can be no doubt that Brazilian steel firms employ more labor in auxiliary activities and office work than American or European firms. Also, as previously noted, many Brazilian steel firms are located in areas where they had to build towns for their workers with all sorts of required social investments. All this has put a substantial cost burden on steel firms.[2] I would argue, however, that although these factors are a substantial cost burden to steel firms, they are not relevant in considering whether the basic productivity position of the firms is high or low. When the pressure of competing against free imports or for exports becomes keen, a society may take various types of actions to eliminate the drag of very expensive social overhead subsidies.

A most relevant measure of the efficiency or productivity of a steel operation is the "coke rate" of blast furnaces, the amount of coke consumed per ton of pig iron produced. Since coke is the most important input into the blast furnace (amounting to about 70 percent of the total value of raw material inputs), the object of a steel mill will be to make the consumption of coke per ton of pig iron produced as small as possible. Various techniques have been invented, or are still being developed, to bring the coke rate down. For example, the coke rate can be decreased by the use of sinter or pellets[3] instead of iron ore, by the injection of oil into the blast furnace or by improving the quality of the coke used. It is thus most important when analyzing the performance of a blast furnace over time to be aware of changes in techniques which

---

picture of the situation" see the article by Erik Ruist, "Comparative Productivity in the Steel Industry" in *Labor Productivity,* edited by John T. Dunlop and Vasilii P. Diatchenko, pp. 175–176.

2. For example, the firms of Volta Redonda, Usiminas, Belgo-Mineira, Acesita, and others are all located in areas where nothing existed before the establishment of these firms. Given Bazil's social legislation and political pressures, each firm had to build substantial social overhead facilities. At Volta Redonda, for example, the rents paid by workers in company houses are purely nominal, and most of the housing expenses are subsidized. These subsidies are not counterbalanced by lower relative wages paid to steel workers.

3. Sinter is a product which results from the fusion of fine iron ores and fine iron-bearing particles with powdered coal and some other fluxes. Pellets are also agglomerates which are hardened after being formed into balls; they are used especially where ores are extremely fine. The coke rate can also be brought down by reducing the size of the particles in the solid charge.

decreased coke rates. Table 29 lists the coke rates of some selected Brazilian and other Latin American firms. Of course, the Belgo-Mineira and the Ferro Brasileiro firms operate charcoal furnaces, and what is of interest here is the change over time in the "charcoal rate." In examining the table, one should know that the most efficient blast furnaces operate in Japan, where a coke rate of 450 kgs. per ton of pig iron has already been attained, whereas in the United States coke rates of 550 to 570 kgs. are presently the norm.

When examining Brazilian coke rates, it should be kept in mind that coke-based steel firms were until recently forced to use 40 percent of domestic coal. Given the poor quality of this coal, its use has been prejudicial to the achievement of high productivity rates in blast furnaces. Technical studies have estimated that Brazilian coke rates could be brought down to levels of about 520 to 550 kgs. of coke per ton of pig iron if imported coal were used exclusively.[4] That is, given the excellent quality of Brazil's iron ore and the techniques of production used currently in Brazil, the country could lower its coke rate to levels below those attained in the United States and many European countries.

Table 29 shows that Brazil has a satisfactory coke rate compared with other Latin American plants.[5] In the case of Brazilian plants, especially Volta Redonda, the decline in the coke rate can be attributed to a large extent to the increasing use of sinter. In the mid-sixties, the cost of coke to Volta Redonda was more than US$ 20.00 per ton, while sinter cost a little more than US$ 3.00 per ton. Increasing use of sinter, with reduction in coke and iron ore inputs thus made substantial cost reduction possible.

The substantial decline in the charcoal rates of Belgo-Mineira and the Cia. Ferro Brasileiro result also from the increasing use of sinter. With the decline of the use of domestic coal and the injection of oil into blast furnaces, Brazil could easily be among the most efficient pig iron producers in the world. Since liquid pig iron is one of the major inputs into the steel furnaces, a low coke rate obviously will make substantial reductions in the cost of steel ingot production; this produces in turn substantial savings in the rolling mill sections.

4. Amaro Lanari Junior, "Consumo de Carvão Nacional Na Siderúrgia," in *Metalurgia*, XXI, No. 93 (Agôsto 1965), 646. Another Brazilian steel expert has told me that according to his calculations, Brazilian firms could even bring their coke rates down to 500 kgs.

5. The coke rate in Indian Steel Mills in 1960 was approximately 900 kgs. per ton; see William A. Johnson, *The Steel Industry of India*, p. 206 *n*.

Table 29

Blast Furance Productivity

(consumption-kgs/ton)

			Iron ore	Sinter	Coke
Brazil					
Volta Redonda		1960	1,416	101	815
		1964	911	635	656
Usiminas		1964	102	1,495	592
	April	1966	334	1,205	623
Chile					
Huachipato		1963	1,613		601
Argentina					
San Nicolas		1963	1,570		700
Mexico					
Monclova		1963	1,060		830

Brazil—charcoal furnaces					Charcoal (m³/ton)	
Belgo-Mineira						
at Monlevade	1940	1,592		4.1	(943)[a]
	1956	200	1,369	3.2	(736)
	1964	844	626	3.0	(690)
Cia. Ferro Brasileiro	1939	1,829		4.9	(1,127)
	1955	1,629		4.0	(920)
	1964	801	820	3.7	(851)

Sources: Calculated from materials made available directly by firms; for non-Brazilian firms, see CEPAL, *La Economia Siderúrgica de America Latina,* mimeographed, Febrero 1966.

a. Numbers in parentheses are estimates of equivalents in coke—multiplying the cubic meters of charcoal by 230 kgs. Comparison is meaningless for cost purposes, however.

ECONOMIES OF SCALE IN INTEGRATED STEEL MILLS

The importance of economies of scale in an integrated steel mill may be understood by an examination of Tables 30 and 31. These were estimates made by the U.N. Economic Commission for Latin America for hypothetical Latin American plants. It will be noted that installed blast furnace capacity per ton of pig iron falls by about 35 percent when comparing blast furnace plants of 400 thousand- and 1.5 million–ton capacity per year; for an open hearth (SM) steel plant the cost of installed capacity per ton of steel ingot falls by 48 percent; and for flat rolled products the cost of installed rolling mill capacity per ton

Table 30

Economies of Scale in Steel: Installed Production Capacity

(dollars of installed capacity per ton)

	Annual capacity in thousands of tons							
	100	200	400	500	800	1000	1500
Blast Furnace: investment per ton of pig iron	94.73	86.07	75.00	69.33	57.53	52.60	48.33
Open-hearth (SM) steel furnace: investment per ton of steel ingot	74.60	69.07	59.07	53.13	42.73	37.27	30.47
Flat-products rolling mill: investment per ton	482.07	427.20	328.67	286.07	219.19	198.33	191.87
Total investment per ton of output	724.15	635.42	484.51	428.76	330.04	314.38	281.86

Source: Comision Economico Para America Latina, Naciones Unidas, *La Economia Siderúrgica de America Latina*, mimeographed, Febrero de 1966.

These estimates were made taking international equipment costs, which were increased by 20 percent to take into account greater transport and assembly cost of equipment in Latin American countries as opposed to more industrialized countries.

falls by 42 percent. These scale economies reduce cost of production per ton of pig iron, steel ingot and flat rolled products (again when increasing output from 400 thousand to 1.5 million tons) by 12.8, 16.5, and 28.7 percent respectively. Economies of scale for nonflat rolling mills are not quite as striking as economies in flat products; however, the capital cost per ton in these mills is also substantially lower, about one third of the capital cost per ton in flat-products mills.

A glance at Table 22 in the previous chapter should make it obvious that by the mid-sixties only Volta Redonda had reached production levels at which scale economies were making themselves felt. Of course, one cannot exactly apply the estimated scale economies of Tables 30 and 31 to individual Brazilian plants. The scale economies estimates were made for a flat-products plant. Its total rolled products output in 1966 was 957,364, with flat products amounting to about two thirds of this total. Thus, Volta Redonda's scale economies in its rolling mills was probably much smaller than those indicated in Tables 30 and 31 for production levels of 1 million tons. Volta Redonda's expansion program could, however, increase its production to levels of 2.5 million in the mid-1970s and the levels of output of Cosipa and Usiminas, specializing in flat products, should pass the one million mark in the

Table 31

Economies of Scale in Steel: Hypothetical Production Costs
in Intergrated Firms Producing Flat Products

(dollar cost per ton)

Annual capacity in thousands of tons

	100	200	400	500	800	1,000	1,500
Pig iron (using 20 percent sinter)	55.04	49.52	45.38	43.88	41.83	40.83	39.57
Steel ingot (SM with oxygen injection)	95.22	87.31	77.89	74.70	70.22	68.02	64.97
Flat Rolled Product	235.49	212.58	169.55	158.29	135.05	126.74	120.85

Source: Same as Table 29.

early seventies. Thus, by the next decade Brazil's largest mills should be producing at levels at which they could benefit from substantial amounts of scale economies.

The production level of Belgo-Mineira in Table 22 should be analyzed with proper qualifications. Although pig iron production was only slightly above the 400,000-ton level, produced in two separate places, this production was based on charcoal. It was mentioned above that charcoal blast furnaces involve about half of the investment cost of coke-based blast furnaces. In addition, until the mid-sixties charcoal was cheaper than coke. Since pig iron production at Belgo-Mineira is cheaper than at a coke-based plant and since pig iron is the main input into the steel furnace, the smaller scale economies in the steel ship of Belgo-Mineira were more than made up by the lower priced input. In the rolling mill section, Belgo-Mineira produces only a small number of flat products (mainly of the type which do not necessitate the same scale as the products of Usiminas or Volta Redonda for economical production) ; most of its output consists of non-flat products.[6] Since scale economies for nonflats are already quite satisfactory at production levels of 300,000 to 400,000 tons, one may assume that the firm was producing at a relatively efficient level.

Most other Brazilian steel firms produce nonflat products or special steels, where scale does not play the role it does in flat products. It is quite evident, however, that higher levels of production of finished rolled products at plants like Mannesmann, Acesita, would still reduce substantially the fixed costs per ton of output.

Part a. of Table 32 gives an idea of the firm size distribution in the steel industry in 1966, and part b. of the table shows the capacity levels which some of the major Brazil firms hope to achieve by the first half of the 1970s. Comparing these data with our previous discussion of scale economies, there is a clear indication that a larger proportion of Brazil's steel firms will reach production levels where they will be able to fully benefit from scale economies.[7]

Cost of Production Estimates

The cost data I shall examine in this section are based on three different estimates. I have made some estimates of my own based on information I obtained directly from firms. Also presented are cost

6. Belgo-Mineira uses a Steckel mill, which is a hot reversing sheet mill adequate for the production of flat products on a small scale. Investment in such a mill is much lower than a regular flat-products mill. The Belgo-Mineira Steckel mill has an annual capacity of 120,000 tons.

7. Not all of the smaller firms which will continue to produce below 100,000 tons are necessarily uneconomic, especially those producing special steels or those using electric furnaces.

Table 32

a. Firm Size Distribution in 1966 According to Levels of Production

(figures in parenthesis refer to absolute number of firms in each size category)

	Pig iron		Steel ingot		Rolled products	
	Yearly Output		Yearly Output		Yearly Output	
	(1000 tons)	percentage	(1000 tons)	percentage	(1000 tons)	percentage
	0-50	13.3 (5)[a]	0-50	11.0 (21)	0-50	10.0 (18)
	50-100	11.6 (5)	50-100	6.9 (3)	50-100	15.7 (6)
	100-500	28.0 (2)	100-400	10.8 (3)	100-250	5.7 (1)
	500-1,500	47.1 (2)	400-500	24.0 (2)	250-500	35.4 (3)
			500-1,000	14.1 (1)	900-1,000	33.2 (1)
			1,000-1,500	33.2 (1)		
	Total	100.0	Total	100.0	Total	100.0

b. Expansion Plans of Major Brazilian Steel Firms

(in millions of ingot tons)

Cia. Siderúrgica Nacional (Volta Redonda)	2.5 (3.5)
Usiminas	1.0 (2.0)
Cosipa	1.0 (2.0)
Belgo-Mineira	.5
Acesita	.2
Barra Mansa	.16
Ferro e Aço Vitória	.3

Sources: Computed from data provided by statistics department of CSN; expansion figures from Booz Allen & Hamilton unpublished report.

Note: Expansion plans listed in part b refer to capacity planned for early seventies. The numbers in parenthesis refer to second stage of planned expansion.

a. This number in parenthesis refers to five small firms which also have steel shops and about twenty tiny blast furnace operations in the state of Minas Gerais.

estimates by the Economic Commission for Latin America for Volta
Redonda and for some other Latin American steel firms. Finally, cost
estimates made for the Brazilian Development Bank (BNDE) and the
World Bank (IBRD) by the international consulting firm Booz Allen
& Hamilton International (BAHINT) are presented separately, since
they are not precisely comparable to the type of cost estimates made by
ECLA and myself. Detailed information on the latter cost estimates
may be found in Appendix III.

Because of the confidential nature of the information supplied to
me by many Brazilian firms, it was not possible to present direct
estimates for individual firms. Instead I have divided Brazil's steel
industry into a number of "representative firms." For example, in the
case of the cost of pig iron production have pooled information on
blast furnace productivity (coke, iron ore, and/or sinter per ton of
pig iron), labor input for a number of large firms, estimating cost
ranges per ton of pig iron. The same was done for medium-sized blast
furnaces using charcoal. In the case of steel shops, information from a
number of firms was obtained to make estimates of the cost of steel
ingot produced by large open hearth (SM) furnace operations, medium-
sized open hearth firms, and large LD-using firms.

Table 33 contains my own cost estimates for various types of Brazilian steel firms and the cost estimates of ECLA for Volta Redonda
and a selected number of other Latin American firms. These estimates
should be presented with a series of qualifications. All my cost estimates
and probably those of ECLA were reached in an indirect way. Information was usually obtained about direct physical raw material and labor
inputs. Sometimes prices of inputs were furnished directly by the firms
and at other times these had to be obtained from reports in trade
journals. Capital costs are, as usual, the most tricky part of cost
estimation. In my own calculations I have used capital charges estimated
by ECLA for Latin American plants, with special adjustments for
Brazilian firms.[8] Taking into account the necessary excess capacity
which an integrated steel mill will have for a period of years after its
inauguration, the capital costs should probably be higher than the ones
actually imputed in my calculations. The rolling mill section is, as

8. The capital charge is based on estimates to be found in "The Iron and Steel
Industry of Latin America: Plans and Perspectives," in *Proceedings: United Nations Interregional Symposium on the Aplication of Modern Technical Practices in
the Iron and Steel Industry to Developing Countries*, pp. 95–120. See Appendix III.

Table 33

Cost Estimates for Brazilian and Other Latin American Iron and Steel Firms

(US$ per ton)

	Brazil 1964	Brazil (Volta Redonda)	CEPAL estimates 1963 Argentina (San Nicolas)	Chile (Huachipato)	Mexico (Monclova)
Blast Furnace (pig iron)					
Large integrated (coke-based)[a]	48.49–50.32	40.50	56.72	45.96	43.11
Medium (charcoal-using)[a]	34.49–38.93				
Steel Furnace					
Large SM	67.36–71.85	71.75	85.37	73.91	71.23
Medium SM	65.35–71.03				
Large LD	68.85–73.44				
Medium LD	60.86–64.92				
Medium electric	64.95–72.75				
Rolling Mills					
Large flat	120.63	156.24	180.99	187.46	172.67
Large nonflat	100.98	115.34		118.70	
Medium nonflat	103.77				
Medium flat	147.27				

Sources: First column estimates based on direct information obtained from various steel firms. Cepal estimates same source as Table 29.

Note: See Appendix III for explanation of special circumstances making LD steel more expensive than SM steel.

a. "Large" or "medium" refers to large or medium integrated firms, except for medium electric steel furnace. When two numbers are cited in first column, they refer to cost ranges, i.e., the highest and the lowest cost estimated.

mentioned before, usually the lumpiest part of the investment; it is rarely used at full capacity in the initial stages of production of a firm. Although fixed capital costs will thus be rather high in the beginning, I have chosen not to take this into account. The main reason for this is that I am interested in the ultimate feasibility or comparative advantage of producing steel in the country. If the market is large enough and growing rapidly enough, the problem of initial high fixed capital costs will take care of itself, that is, disappear.

It will be noted in Table 33 that except for my estimates of the cost of pig iron production Brazil has lower costs than other Latin American plants. One should note especially the low cost of pig iron production of medium firms, since they all utilize charcoal.

For comparative purposes it would be useful to mention direct-cost information I received from one of Brazil's major steel firms for April 1966. Accordingly, the cost of one ton of liquid pig iron was US$44.38, of one ton of steel ingot was US$53.50, and of one ton of slabs US$63.07. These costs were net of capital charges. Had these been included, the total cost per ton of output would have been in the lower ranges of my own estimates. Finally, the estimates made by the consulting firm Booz Allen & Hamilton International for the BNDE and the World Bank have costs of pig iron varying between US$31.50 and US$40.00 for large-scale integrated mills and costs as low as US$23.60 for charcoal-based pig iron.

A brief comment is needed in connection with my cost estimates of ingot steel produced by open hearth steel shops (SM) and LD converters. The advantages of the latter, and the reason for its increased utilization in newer steel works, are the much smaller capital cost involved.[9] Depending on the size of the operation, the capital cost of an LD plant can vary from one half to one third of the capital cost of an open hearth steel shop. Usually the low capital cost of an LD shop more than overcomes the higher cost of the metallics it needs, especially more liquid iron per ton of ingot steel produced than an SM furnace. Also, general operating costs are lower for an LD converter than an open hearth operation. As will be seen in Appendix III, I have adjusted capital consumption for the LD cost estimates. The cost of steel ingot

9. Walter Adams, and Joel B. Dirlam, "Big Steel, Invention, and Innovation," *The Quarterly Journal of Economics* (May 1966), p. 179. See G. S. Maddala and Peter T. Knight, "International Diffusion of Technical Change—A Case Study of the Oxygen Steel Making Process," *The Economic Journal*, September 1967.

of large LD-using plants is not lower than SM steel ingots, however, because of the substantially higher consumption of liquid pig iron, I assumed, and because of possibly too high a blow-up of my data for "other conversion costs." My prime sources of data were LD shops which were fairly new and possibly had not yet reached a peak of efficiency.

Tables 34 and 35 contain my own and ECLA's estimates of the basic cost structure per ton of output in each of the main sections of the steel firm. Detailed data on which these breakdowns are based may be found in Appendix III. The ECLA estimates refer to a number of steel mills throughout Latin America, Volta Redonda being the only Brazilian firm included. The basic difference between the ECLA estimates and my own is that raw material input of the ECLA estimates is much smaller than mine, while the capital proportion of ECLA is larger. In the case of Volta Redonda, ECLA's capital charges are twice to three times as high as the one I have used in my estimates for large scale plants in Brazil. There are a number of reasons for this. First, the production level for the ECLA calculation was smaller than the level I assumed for my 1964 estimates of large plants; they thus have a higher fixed capital cost per unit of output. Second, the ECLA calculations included amortization and interest payments, which are not included in the table of capital costs I used. The ECLA capital charges I used were based simply on 9 percent of invested capital in each section of the steel mill.

Since my estimates were based on selected information obtained directly from firms, I was not in a position to obtain the type of information contained in the ECLA estimates which fall under the heading of "other conversion costs."[10] I got around this problem by blowing up my estimates in accordance with the recommendation of some engineers. For example, the estimate I obtained for the cost of pig iron was based on information which covered about 90 percent of costs. I thus increased my estimates accordingly. This does not, however, affect the proportions in Table 34. The above qualifications to the information presented in Tables 34 and 35 should make it clear that a comparison of the tables is not very profitable. The information of Table 34 is of interest because of the comparison one gets among various types of

---

10. "Other conversion costs" cover such items as repairs and water in the case of a blast furnace; fuels, refractory bricks, general services for a steel shop; refractories and services for the rolling mills.

steel operations within Brazil, while Table 35 is of some value because of the international comparisons one can make. That is, comparisons contained within the tables are valid since the estimates are based on comparable information.

Let us first account for some of the differences in the cost estimation of ECLA and my own in Table 33. In all probability, my blast furnace cost estimates are somewhat high. Although the capital charge for Volta Redonda used by ECLA was substantially above the one I used, ECLA's estimates of raw material costs and of "other conversion costs" were considerably below my own (all this is discussed in Appendix III). ECLA also "credited" the raw material inputs for gas—the value of the amount of gas produced per ton of pig iron was subtracted from total raw material inputs into the blast furnace—which I did not do with my estimates. All this brought the ECLA estimates below my own. The steel shop cost estimates are closer. Here I assumed a more efficient use of pig iron (lower pig iron use per ton of steel ingot produced) than ECLA and, of course, a lower capital cost. Most of the difference in the cost of rolled products lies in the substantially higher capital charge used by ECLA.

Turning to Table 34, it is interesting to note the relatively low labor input requirements in the larger steel operations. The larger labor input of the smaller charcoal blast furnace operations and of the smaller rolling mills reflect the more simple technology used in such enterprises (see discussion in Chapter Two). The comparisons in Table 35 are of interest because they reflect to a certain extent the relative factor endowments of each country. Brazil's lower proportions of iron ore and sinter cost in the blast furnace section compared with other countries obviously reflects the country's abundant endowment of cheap high-grade iron ore. Brazil's high proportion of coke costs compared with the other countries reflects both the country's great dependence on imported coal, and it especially reflects the increased costs brought about by the required 40 percent input of domestic coal.

Another analysis of the cost structure of Brazil's steel industry was made by the consulting firm Booz Allen & Hamilton International, which had access to the books of the sixteen most important firms for the period October-November 1965. Since the information obtained was of a confidential nature, I was only able to reproduce aggregate data. This is done in part a of Table 36. I have, however, included in parenthesis the range of the sixteen observations for each cost item.

Table 34

Cost Structure of Brazilian Steel Output

(percentage distribution)

a. *Blast furnaces* (cost structure per ton of pig iron)

	Large coke-based Blast furnaces		Medium sized charcoal-using	Small charcoal-using
	a	b		
Raw materials	87.7	86.3	85.4	78.2
Manpower	1.8	2.9	3.6	10.3
Capital charge	10.5	10.8	11.0	11.5
	100.0	100.0	100.0	100.0

b. *Steel shops* (cost structure per ton of ingot steel)

	Large SM	Large LD	Medium LD
Raw materials[a]	89.9	93.3	89.4
Manpower	3.4	2.0	2.9
Capital	6.7	4.7	7.7
	100.0	100.0	100.0

c. *Rolling mills* (cost structure per ton of product)

	Large flat[c]		Large nonflat[c]		Medium flat	Medium nonflat
Raw materials[b]	74.9	(80.6)	73.2	(80.6)	63.7	79.7
Manpower	0.7	(0.8)	0.8	(0.9)	4.5	6.1
Electric energy	2.1	(2.3)	2.6	(2.8)	2.3	3.3
Capital	15.2	(16.3)	14.3	(15.7)	29.5	10.9
10% conversion cost	7.1		9.1			
	100.0	(100.0)	100.0	(100.0)	100.0	100.0

Source: Information obtained directly from various steel firms.

a. Largest portion is liquid pig iron.

b. Mainly input of ingots.

c. Numbers in parentheses are proportions based on total not containing conversion costs in order to make comparable with estimates for medium-sized firms.

It will be noted that fixed cost as a proportion of total cost was much higher in the flat-products firms than in the other sectors. The reason in part for this is the more complex installations which a large-scale flat-products producer requires and in part because many large-scale flat producers that entered the market only in the sixties were still producing substantially below capacity. Thus older producers of flat products had fixed costs amounting to only 24 percent of revenue, while newer producers had fixed costs as high as 31 percent of revenue.

Table 35

International Comparison of Steel Cost Structure

(percentage distribution)

a. *Blast furnace* (cost structure per ton of pig iron)

	Brazil	Argentina	Chile	Mexico
	(Volta Redonda)	(San Nicolas)	(Huachipato)	(Monclova)
Raw materials	60.4	73.7	61.5	55.2
(Coke)	(49.2)	(33.4)	(38.7)	(25.1)
(Iron ore)	(4.5)	(40.3)	(22.8)	(17.4)
(Sinter)	(6.7)			(12.7)
Labor	1.8	1.7	3.6	2.7
Other conversion[a]	7.5	5.2	8.9	12.0
Capital charge	30.3	19.4	26.0	30.1
Total	100.0	100.0	100.0	100.0

b. *Steel shop*

Raw materials[b]	66.2	73.0	65.2	67.4
Labor	2.4	2.1	4.8	2.8
Other conversion[a]	20.1	14.7	17.4	18.9
Capital charge	11.3	10.2	12.6	10.9
Total	100.0	100.0	100.0	100.0

c. *Flat rolling mill*

Raw materials[b]	61.7	56.3	55.4	57.9
(Ingots)[b]	(68.1)	(63.0)	(63.5)	(64.8)
Labor	1.9	1.1	4.4	1.4
Other conversion[a]	6.1	4.6	5.9	6.3
Capital charge	30.3	38.0	34.3	34.4
Total	100.0	100.0	100.0	100.0

d. *Nonflat rolling mill*

Raw materials[b]	73.5		72.9	
(Ingots)[b]	(78.2)		(78.7)	
Labor	1.2		3.1	
Other conversion[a]	7.9		7.7	
Capital charge	17.4		16.3	
Total	100.0		100.0	

Source: Calculated from same source as Table 29.

a. Other conversion charges include refractory bricks, fuels, electric energy, repairs, etc.

b. Raw material input in steel shop means principally liquid and solid pig iron, scrap, and assorted minerals; in rolling mill raw material input means mainly ingot and combustibles (ingot input is larger than raw material inputs into rolling mills because of a credit being given for the scrap recovered in rolling the ingots).

Administrative costs were relatively low, but it was never stated what part of administrative costs were included under fixed costs.

Of special interest in Table 36a is the column indicating profits after taxes as a proportion of sales. This is also compared with the range found for some of the larger United States steel firms. It will be noted that the performance according to this criterion compares rather favorably with the United States. A number of qualifications are in order, however. First, the Brazilian data are for 1965, a year of recession during which many steel products were sold domestically and abroad below cost. Second, many of the firms producing at a loss, especially in the production of flat products, had recently begun to operate and were not yet producing at a breakeven level. Most of the firms that were well established had profit rates that compared favorably with those in the United States. Since Booz Allen & Hamilton also made an estimate of the value of net facilities for all firms it studied, I made an estimate of the rate of return on invested capital for two of the larger firms that have been established for a longer period of time. That is, I took the ratio of profits after taxes to the value of the net facilities. I obtained the rates of return of 5.4 and 14.3 percent. It should again be emphasized that these were rates during a year of recession, when the inflation continued but steel prices were kept artificially behind price increases. One would thus suppose that in a more nearly normal year these rates would be substantially higher, possibly ranging from 10 to 20 percent. Profits after taxes as a percentage of invested capital for a sample of United States steel firms in 1965 varied between 7.6 and 12.5 percent.[11]

11. A calculation of the ratio of net profits to the value of total equipment and installations shows Volta Redonda (CSN) to be in a position superior to American firms. The calculation gave the following results:

Volta Redonda	1963—29%
	1964—27%
	1965—22%
U.S. Steel	1965—10%
Bethelehem Steel	1965—11%
Republic Steel	1965—10%
National Steel	1965—18%
Youngstown Steel	1965—12%

Calculations made by Aluisio Marins from the following sources: Companhia Siderúrgica Nacional, *Relatorio da Diretoria,* 1965; and *Steel,* April 4, 1966.

There is a possibility that Volta Redonda's profits are somewhat overstated, since they might still contain a certain degree of illusory profits. That is, although the inflation was taken into account in making depreciation calculations, the full impact of the inflation might not have been accounted for.

Table 36

a. Revenue and Cost of Brazilian Steel Firms

(in US$ per ton of finished products)

Group of Firms Producing:[a]

	Flat products		Nonflat products		Special steels[b]	
Revenue	120	(113–127)	111	(100–141)	292	(232–658)
Variable cost	61	(56–63)	61	(54–84)	69	(93–240
Fixed cost	36	(30–45)	18	(7–26)	48	(25–89)
Administrative cost	9	(7–14)	5	(1–13)	14	(11–26)
Interest	11	(5–31)	9	(4–22)	16	(5–41)
Miscellaneous cost	9	(9–10)	8	(2–17)	26	(14–79)
Profit	−3	(−30–+13)	+8	(−11–+30)	+69	(+30–+204)
Profit as percentage of sales[c]	−5	(−26–+9)	+6	(−16–+15)	+19	(+ 5–+29)

Five major United States steel
firms in 1965—
profit as percentage of
of sales   6.7 (5.6−7.9)

Source: Brazilian Steel Industry Survey, by Booz Allen & Hamilton International, Rio de Janeiro: Banco do Desenvolvimento Econômico, August 18, 1966, unpublished report. U. S. profit as percentage of sales figures are taken from *Fortune,* July 15, 1966.

a. These numbers are averages. A firm producing both flat and nonflat products would be placed in one or the other category depending on which type of product predominates. Numbers in parentheses are ranges—lowest to highest.
b. Special steels also include firms producing steel tubes.
c. Profit after income taxes. Numbers in this row are all precentages.

Part b. of Table 36 contains cost-of-production estimates for flat products of Usiminas and similar plants in the United States. According to these data, the cost-of-production position of Brazil is quite favorable in relation to the United States. In connection with these data it should be mentioned that by the first half of 1967 the government had forced steel prices to fall substantially behind the general price increase in Brazil: this had obvious prejudicial effects on the profit position of Usiminas and other firms. It is also interesting to note that both the high external debt of a firm like Usiminas and the nature of Brazilian indirect taxes represented a substantial cost burden for the firm in comparison with firms in the United States.

When examining the cost data in Table 33, one should also keep in mind the handicap under which Brazilian coke-based firms are operat-

Table 36 (Continued)

b. Special United States and Usiminas Cost Estimates for the
First Half of 1967[a]

(in US$ per ton)

	Usiminas	USA
Operational costs	89.86	105.95
(Production costs)	(81.53)	(98.71)
(Administrative and sales costs)	(8.33)	(7.24)
Financial Costs	60.89	9.87
(Interest)	(40.56)	(1.32)
(Depreciation)	(20.33)	(8.55)
Taxes	22.93	14.94
(Income tax)	(—)	(5.92)
(Other taxes)	(22.93)	(1.97)
Profit or loss	−48.60	+56.50
Price of product	125.08	131.61

Source: Printed in *Estado de São Paulo*, July 27, 1967, p. 21. Based on a talk by the president of USIMINAS, Amaro Lanari Junior.

a. Average cost for flat products.

ing with the use of up to 40 percent of low-grade Brazilian coal. It has been estimated that this coal raises the cost of producing pig iron by about US$12 a ton, and that this, in turn, raises the cost of producing steel ingots by about US$11 and the production of rolled products varies between US$14 and US$15 a ton.[12] Even taking into account that these estimates might be a little high, it should be obvious that the exclusive use of imported coal could lower the costs per ton by a substantial amount relative to other Latin American steel producers or even American and European producers. Also, applying these cost reductions to the estimates in Table 33 for large, integrated coke-based plants, the cost advantage of charcoal-using plants almost totally disappears.

A COMPARISON OF STEEL PRICES

Table 37 contains various attempts at international price comparisons of steel products. The table presents a comparison of the factory

12. A. Lanari, "Consumo de Carvão Nacional na Siderúrgia," *Geologia e Metalurgia,* No. 27 (1965), pp. 247–248. Dr. Lanari's calculations are reproduced in Appendix IV.

prices of selected products at Volta Redonda and in the United States in May 1967, shortly after an increase in Volta Redonda's steel prices and two months after a devaluation of the cruzeiro. At that time, Volta Redonda seems to have had some price advantages in the production of rails, heavy plate and hot rolled sheets.

In its study of the Latin American steel industry, ECLA tried among other things to make some international steel price comparisons in order to get some idea of the relative efficiency of steel industries. I have reproduced in Table 38 some of the ECLA estimates of dollar prices for steel products in various countries. Although Brazil turns out in most cases to be the second lowest priced producer in Latin America, these price figures are immediately suspect because in most cases Colombia, which is known to have one of the least efficient steel mills in Latin America, is consistently the lowest priced producer. One obviously runs into the exchange-rate problem: which is the correct one to use at which time?

Table 37

Comparison of Brazilian and United States Factory Prices (May 1967)

	(in NCR\$/t)	
	Volta Redonda	United States
Rails	341	361
Heavy plates	355	394
Hot rolled sheets	366	397
Cold rolled sheets	475	437
Galvanized sheets	637	558
Tin plate	638	508

Sources: Volta Redonda Prices furnished directly; United States Prices from *Steel*, May 1967.
Note: United States prices converted at exchange rate US\$1 = NCR\$2,70. Products compared have same specifications. Heavy Plate price is of Usiminas. These data were kindly furnished to me by G. Aluisio Márins.

The variations one obtains in making conversions of prices in local currency into dollar prices are obvious when examining Table 38b. Here I have converted internal prices at various periods of time using the prevailing exchange rates. One should note especially the prices in October 1965 and March 1967. The former was the month before a devaluation from Cr\$ 1,850 to Cr\$ 2,200 to the dollar and the latter was the month in which the cruzeiro was devaluated from Cr\$ 2,200 to 2,700 to the dollar. I used the old and the new exchange rates for each month. For October 1965, I used the old one and the one instituted in November 1965, and for March 1967 I used the new Cr$ 2,700 rate and the rate of Cr$ 3,000. The latter was tried, since it has been claimed that the March 1967 devaluation was not large enough compared with changes in the domestic price level. Table 38 clearly reveals how sensitive prices are to the exchange rate one chooses and how close at times the Brazilian price comes to European prices. Table 38c contains the price ranges of October-November 1965 for some products by a special study group which obtained direct price information from firms. These prices are possibly a little on the low side because they might contain special reductions offered to customers, but not listed, and they are prices at the mill rather than at the places of consumption.

The evidence presented indicates that Brazil's prices are among the lowest in Latin America and, depending on which exchange rate one chooses, are not necessarily out of line compared with European prices. Taking into account the cost reductions that could be attained by the increasing or exclusive use of imported coal and the economies of scale that can be attained as output increases, one is led to the conclusion that Brazil seems to be in a favorable cost position in the production of steel.

PROTECTION AND PRICE FORMATION

The Brazilian steel industry has benefited from tariff protection and, at times, from other institutional arrangement. In the case of flat products, Volta Redonda once had a monopoly position as the country's sole importer. It imported products not produced in Brazil or it imported products produced domestically when shortages appeared at prevailing domestic prices. Volta Redonda was also the sole importer of nonflat products, but throughout most of the post–World War II period the country has been self-sufficient in such products as bars, light sections tubes, and rails. In the case of special steels, ACESITA once had a role similar to that of Volta Redonda as import monopolist. Firms wanting to import special steel products had to prove to the Banco do Brasil that there were no domestic producers capable of supplying the needed steel.[13]

Brazilian steel tariff levels have always been high enough to protect

13. They had to present letters from special steel producers stating that they were in no position to supply the product wanted.

Table 38

a. ECLA Comparison of Steel Prices in Various Latin American and European Countries

(Price in US$ per ton) (prices for August 1965)

	Argentina	Brazil	Colombia	Chile	Mexico	Peru	Venezuela	West Germany	Belgium	France
Concrete rods	238	139	122	164	154	206	148	108	99	104
Cold rolled sheets	287	243	165	298	208			158	153	149

Source: CEPAL, *La Economía Siderúrgica de América Latina,* February 1966, mimeographed, p. 170.

b. Brazilian Steel Price Estimates at Various Time Periods

(Price in US$ per ton)

	October 1964	June '64	Jan '65	Oct. '65		Oct. '66	March '67	
				A	B		C	D
Round bars	230–270	190–230	190–220	130–150	110–130	220–240	170–230	150–210
Cold rolled sheets	250–310	220–280	210–270					

Source: Calculated from prices listed in *Máquinas & Metais* of CSN; in all cases exchange rate used was the prevailing one, except in October 1965 when both the Cr$ 1,850 (A) and the Cr$ 2,200 (B) were used and in March 1967 when the exchange rate Cr$ 2,700 (C) and Cr$ 3,000 (D) to the dollar were used. Price ranges are used, representing the cheapest and the most expensive product in each category.

c. BNDE–World Bank–Booz Allen Price Estimates (October–November 1965)

(Price in US$ per ton)

Concrete rod	—89–127
Rolled bars	—94–142
Cold rolled sheets	—98–145

Source: Unpublished study for BNDE and World Bank by Booz Allen; prices were estimated for various firms—ranges represent cheapest and dearest prices. Rolled bars exclude bars of special steels.

effectively domestic production. Tariffs are based on the CIF price of the product. From 1957 until early in 1967, rates for rails ranged from 30 to 60 percent; for other nonflat products, 50 to 60 percent; for heavy plates, 50 to 60 percent; for other flat products, 20 to 50 percent. In March 1967, tariffs were lowered. For both nonflat and flat products, however, they still ranged between 15 and 40 percent.

Until the mid-1960s, price-setting for flat products was dominated by Volta Redonda. Until the sixties, Volta Redonda and Belgo-Mineira controlled the domestic market. Given the former's size, it was always the price leader. Belgo-Mineira never really was in a position to threaten this leadership, but it was also not interested in challenging it. Since until the sixties steel was a sellers' market, and since Volta Redonda was a higher cost producer than Belgo-Mineira, the latter only too willingly accepted the former's leadership. Since Volta Redonda is a government-controlled firm, its price-setting policies are often influenced by broader economic policies of the government rather than by market and cost conditions. On a number of occasions, the government has ordered Volta Redonda to set its prices in accordance with its price stabilization objectives, regardless of what had happened to the company's costs.[14] Such direct government interference in pricing was evident even in the mid-sixties, after Usiminas and Cosipa had entered the market. In 1965, the Brazilian government passed a decree giving firms certain tax advantages if they would not raise their prices by more than 10 percent a year.[15] The government-controlled steel firms

14. For an interesting account of political interference with steel pricing in 1963 see Frank P. Sherwood, *O Aumento do Preço do Aço da C. S. N.—Estudo de Um Caso,* Rio de Janeiro: Fundação Getúlio Vargas, Cadernos de Administração Pública —61, 1966.

15. This decree was known as Portaria 71. It brought many firms into difficulties, since prices of inputs rose substantially more than the allowed price increase of steel products of 10 percent. A comparison of the evolution of Volta Redonda's product prices, its costs of production and the changes in the country's general wholesale prices (excluding coffee) will make the difficulties of steel producers clear. For these comparisons, January 1965 is taken as the base, except for cost of production where January and February are taken as the base.

	Volta Redonda's Cost of Production	Prices of VR's products	Wholesale prices (excluding coffee)
1965—January	99.3	100.0	100.0
February	100.7	100.0	101.8
December	124.4	100.0	125.1
1966—January	130.9	110.0	136.5
October	151.3	110.0	174.2
November	153.7	121.0	181.1

were forced to join this commitment, and so also were private firms, since most of them relied on government credit for their expansion programs.

In many nonflat steel products a greater degree of price competition existed among producers than in flat products. It is difficult, however, to determine the degree of this competition. Many products by private producers are officially sold at the listed prices of Volta Redonda, but actually all sorts of price concessions are granted which are almost impossible to identify. Many steel products are also sold by middlemen, whose speculation has caused substantial price fluctuation. These middlemen have proved difficult to eliminate, since they possess a credit mechanism that most steel firms were not in a position to offer.

QUALITY OF BRAZILIAN STEEL

If the quantitative evidence presented above is sufficient to convince the reader that Brazil has been developing into an efficient and relatively low-cost steel producer, there might still remain some doubts about the quality of the product. The quality of Brazil's iron and steel has been a function of the increasing complexity of Brazil's industrial structure. So long as the construction and railroad sectors were the main customers for steel products in Brazil, the need for rigorous quality control was not very pressing. With the rise of such industries as automobile manufacturing and naval construction, however, which because of government policies were forced to rely on domestic inputs, there arose a need for substantially increasing quality control. In the late fifties, when the automobile sector became a major customer of the Brazilian steel industry, the rejection rate of steel products was very high. The types of defects the automobile producers encountered in the steel products were badly rolled products with unsatisfactory surfaces, bars or sheets sent with wrong dimensions; often the quality of the steel itself was defective.[16]

1967—January	156.0	121.0	181.1
March	—	129.5	189.3

Source: Information obtained from CSN; wholesale prices are from Fundação Getúlio Vargas.

16. For a more thorough discussion of the steel quality problems in the late fifties, see José Bento Hucke and Domingos Espósito Neto, "Defeitos de Aços Nacionais Para a Indústria Automobilística," *ABM, Boletim da Associacão Brasileira De Metais,* (Outubro 1960), pp. 741–749.

The high rejection rate and the pressure from customers have forced the industry to improve substantially the quality of its product. It is thus interesting to note that because of the diversification of Brazil's industrial complex the country benefitted from an external economy in the sense that this diversification resulted in a substantial product improvement.

7

PERFORMANCE OF BRAZIL'S
STEEL INDUSTRY: EXTERNAL

As MAY be seen in Table 39, more than 70 percent of Brazil's apparent consumption of steel takes place in and around the cities of São Paulo and Rio de Janeiro. Although greater São Paulo's industrial complex absorbed almost 50 percent of the country's steel production, it accounted for less than 20 percent of steel production in the mid-sixties (compare Tables 39 and 40). On the other hand, the state of Minas Gerais, which produced on the average about 35 percent of the country's steel in the mid-sixties, consumed less than 10 percent of the output. As a rough generalization, we can thus say that less than one fifth of Brazil's steel production was market-oriented in the mid-sixties, about a third was resource-oriented (Minas Gerais output taking place close to the iron ore mines), and roughly 40 percent was "half market-oriented"—the production of steel at Volta Redonda and some other smaller firms in the state of Rio de Janeiro (like the firm Barra Mansa), which takes place at a location halfway between raw material source and market.

An analysis of the geographical distribution of steel-making capacity shows that there was a greater market-orientation in capacity than in the production pattern in the mid-sixties. The production facilities are about evenly divided between market-orientation, "in-between" location, and resource-orientation. The greater weight of the market in production capacity is the result of the new facilities of Cosipa (a large proportion of which are still not fully used) and of such special steel firms as Aços Anhanguera, which only began operating in 1966.

Until the construction of Volta Redonda, no effort was made to study various alternative sites before constructing iron- and steel-producing concerns. As noted in Chapter Four, the firms that settled in the São Paulo area were often created as extensions to industrial enterprises that wanted to produce their own special steels. In other words, they were the result of backward integration. The firms that sprang up in Minas Gerais were created by mine owners anxious to "industrialize" their raw materials. It was also seen that in the case of the creation of the firm Belgo-Mineira, the government of the state of Minas Gerais had a strong promotional hand.

The actual, as opposed to the published, reasons for the location of Volta Redonda in the Paraiba valley between the biggest consuming centers and about one third of the way to the iron mines has been shrouded in controversy. Some say that the location was the result of political compromise between the claims of the urban consuming centers and the state of Minas Gerais. Others say that the site was ideal— between the two main consuming centers, with a favorable railway link to the iron mines and not too far from the port of Rio de Janeiro from which the imported coal would be shipped. Modern steel location analysts recommend tidewater plants for firms that rely on the import of one of the principal raw material inputs. Volta Redonda was built during World War II, however, when the Brazilian military were hesitant about locating a steel mill within easy reach of potential enemy ships. Still others mention the fact that the locational decision of Volta Redonda was also influenced by the government's desire to assist the economically rundown Paraiba valley, with an abundant but badly utilized labor force. Also, the Volta Redonda location fulfilled the technical needs of a large integrated steel enterprise: location next to a large and steady supply of water, closeness to abundant power supplies, proximity to an adequate transportation system,[1] and the availability of ample open space for an adequate plant layout with room for expansion.

The planners and builders of Volta Redonda also possessed a certain zeal for social experimentation. They were anxious to build a new model industrial city around the steel works. Volta Redonda was to symbolize a big step, not only towards the construction of an industrial Brazil, but also towards a modern industrial social system. An elaborate workers' city was built, with health centers, shopping districts, recrea-

1. In the case of Volta Redonda, the mill was located at the Central do Brasil Railroad network, which connected Rio and São Paulo and which also had a line into Minas Gerais close to the iron mines. Thus, investment in a transportation network did not have to be done from scratch; investments needed to serve the steel mill could be built into the existing transport system.

Table 39

Regional Distribution of Apparent Steel Consumption

a. *Consultec estimates for 1961*

North and northeast	6.3%
Central Brazil	11.2
Guanabara and Rio de Janeiro State	28.7
São Paulo	46.9
South	6.9
	100.0

Source: Ministério do Planejamento e Coordenação Economica, EPEA, *Siderurgia Metais Não-Ferrosos*, Diagnóstico Preliminar, Rio de Janeiro, April 1966, p. 44. These estimates were originally made by the firm Consultec.

b. *Regional distribution of sales of selected firms in 1964*
(percentage distribution)

	São Paulo	Guanabara and Rio de Janeiro	Minas Gerais	South	Other	Export	Total
CSN (Volta Redonda)							
Flats	53.8	23.6	2.7	10.0	6.5	3.4	100.0
Nonflats	48.4	26.7	9.2	7.6	7.7	0.4	100.0
Belgo-Mineira Monlevade							
Flats	69.7	9.9	7.7	1.3	0.1	11.3	100.0
Nonflats	13.7	4.5	81.2		0.6		100.0
Mannesmann							
Nonflats	70.3	5.9	0.6	0.9		22.3	100.0
Seamless Tubes	31.3	49.2	6.0	4.9	6.1	0.9	100.0
Usiminas							
Flats	18.7	48.9	7.6	1.8	0.2	22.8	100.0
Acesita							
Flats	86.0	9.0	1.0	3.0	1.0		100.0
Nonflats	63.0	21.0	6.0	6.0	4.0		100.0

Source: Special study of the firm Tecnometal for the BNDE, unpublished.

tion areas, and a school system extending from primary school to a higher engineering faculty. Much of this elaborate social infrastructure substantially increased the cost of establishing and running the enterprise. Of course, the planners can claim that the welfare created by this social infrastructure should be included in the social returns of Volta Redonda, if anyone would be inclined to make a benefit-cost analysis of the enterprise. A substantial amount of criticism, however, has been leveled at the subsidy rates at which Volta Redonda's workers

Table 40

a. Regional Distribution of Steel Production

(percentage distribution)

	Rio de Janeiro (state)	São Paulo	Minas Gerais	Other	Total
1965					
Total production	48.3	13.8	33.8	4.1	100.0
Flats	65.4	9.9	24.7	0.0	100.0
Nonflats	33.2	17.7	40.9	8.2	100.0
1966					
Total production	38.0	18.6	36.0	7.4	100.0

b. Regional Distribution of Steel-Production Capacity

(percentage distribution)

Total capacity	32.5	30.0	32.4	5.1	100.0
Flats	26.5	33.4	35.6	9.5	100.0
Nonflats	44.6	26.4	29.0	—	100.0

Source: Calculated from Booz Allen & Hamilton International study for the World Bank and BNDE and from data furnished by the statistical survey section of the Companhia Siderúrgica Nacional.

receive various services. Most of the public utility rates are extremely low and rent on housing is extremely low because it has not kept up with the inflation. For all practical purposes, housing is free. Taking into account that the wages of workers are not below those of industrial workers in Brazil, real wages (including all the benefits) are thus substantially above average industrial wages in Brazil.

Volta Redonda has also encountered a number of problems not anticipated at the time it was constructed. The city was built in gradations on a number of hills surrounding the plant. The buildings are grouped in such a way that the workers' houses are usually in one area, close to the bottom of the hills. The foremen's houses are a little higher, and those of the engineers are higher still. Higher up, the bureaucrats have still better residences. This gradation is also transmitted into the social life of the city, where everyone is connected with the firm. Some engineers I talked with compared the life at Volta Redonda with that in a military garrison town. In addition, the city was built in such a way that almost everyone can see part of the steel works from his home. These factors have caused psychological problems to many workers and professional people. It has caused many of the professionals to leave the place as soon as possible, either to work in CSN's offices in Rio and other cities or to work elsewhere. To that extent, Volta Redonda has been a sort of training center for engineers and other professionals, many leaving as soon as their experiences at Volta Redonda increased their value in the job market. The COSIPA operation, for instance, has attracted many ex–Volta Redonda people because of its location near a highly urbanized area.

Problems of a similar nature are also encountered, for example, in the Monlevade works of Belgo-Mineira. Although facilities of this private firm are not as elaborate as those at Volta Redonda, Brazil's social legislation has forced the company to put substantial sums into housing, education, recreation, and social activities.

The location of COSIPA and USIMINAS was also determined by a mixture of economic and political considerations. It will be remembered from our historical chapter that São Paulo industrial groups had been pressing for an integrated steel plant located in the greater São Paulo area ever since the early fifties. The interest in a large integrated steel mill in the Vale do Rio Doce area, which would be complementary to the iron ore exporting operations of the region go back to the second decade of the century when Percival Farqhuar was trying to push through his schemes. Thus, given the post–World War II needs for further expansion of Brazil's steel-producing capacity, the economic and political interests of São Paulo and Minas Gerais began their pressures for the establishment of large, integrated steel enterprises in their areas.

The interest groups of both regions developed arguments and made studies attempting to prove that theirs was the most logical region for steel investments. Naturally, the Paulistas developed elaborate arguments to show that a market-oriented steel mill made more sense. The site at Santos was close to the principal market; being a tidewater plant[2] it could receive imported coal most cheaply. Domestic coal was shipped up from Santa Catarina. It was also in a good position as an exporter.[3]

The arguments in favor of a resource-based plant in the Vale do

2. Actually, the plant is not exactly in a port, but at an inlet of a small bay.
3. The best-known rationalization for the location of Cosipa can be found in Jorge F. Kafuri and Antonio Dias Leite Jr., "Estudo Da Localizacão De Uma Industria Siderúrgica," *Revista Brasileira de Economia,* Setembro de 1957, pp. 37–78.

Rio Doce region were that it was close to the iron ore mines, there already existed a railroad system serving the Companhia do Vale do Rio Doce in its shipment of iron ore to the port of Vitoria for export, both iron ore ships and railroad cars usually returned empty to their points of origin and it would thus make eminent sense for these to return with coal for a steel plant in the region, there was plenty of space and water and power sites.[4]

As we have already seen, both areas eventually obtained their plants. There can be no doubt that one advantage Cosipa had over the Usiminas works was its closeness to the city of Santos; most workers live in that city and thus it was not necessary for the firm to build a huge social infrastructure. Usiminas, however, was located in a fairly unpopulated area and the firm, like Volta Redonda, was obliged to build a huge social infrastructure (a whole town, with all the facilities necessary under Brazilian social legislation).[5] A number of Brazilian steel experts have stated that Vitória would have been a more logical site for a resources-oriented plant. One might speculate that if Vitória had been a city of the politically powerful state of Minas Gerais, it would have been favored for the project.[6]

The Transport System

Brazil's transport system in the areas relevant to the steel industry is not in condition for efficient use. A large proportion of the rail network was built to serve the rural sector, the tracks often meandering around various properties in a pattern originally designed to pick up rural produce. Thus, for example, the rail mileage from the town of Monlevade to São Paulo is 1,100 kms. compared with 696 kms. by road. Also, rail service in the past has been substantially below internationally accepted standards. These factors have led steel producers to use trucks whenever possible, even though under normal circumstances rail services would have been more efficient for certain key inputs and outputs.

4. The most systematic defense for the Usiminas location can be found in Amaro Lanari Jr., "O Projeto da Usiminas e Sua Justificativa no Planejamento da Siderurgia Brasileira," *Geologia e Metalurgia,* 1961, No. 23.

5. It is interesting to note that the planners of Usiminas tried to avoid some of the mistakes of Volta Redonda, e.g., company housing settlements are located in various areas away from the site of the plant.

6. It is true that the Cia. Ferro e Aço de Vitória was located in that city. Until the mid-sixties the facilities of the company have consisted only of a rolling-mill installation. There are plans to integrate the firm backward and thus transform the company into an integrated steel mill.

Even where railroads have been used, a considerable shortage of rolling stock has in the past existed, forcing many companies to purchase their own cars. Shortage of rolling stock has been one of the principal problems of Cosipa, which relies primarily on railroads for its supply of iron ore.

Maritime transport for the steel industry between Brazilian ports has been generally limited to the transportation of national coal from the port of Imbituba in the state of Santa Catarina to the ports serving integrated steel plants. Utilization of coastal shipping for other inputs or for steel products has been avoided as much as possible because of excessive costs, uncertain and long delivery periods, and poor service in terms of cargo protection. The high costs also result partially from the limited loading and unloading capacity of the ports. These factors were instrumental, for example, in Cosipa's building its own port facilities. The latter, however, were still not ready in 1967, hence the heavy reliance of the company on rail services. The use of big barges (11,000 tons) pulled by a trawler will be initiated in the early months of 1968. If this system proves successful, Cosipa's iron ore and domestic coal transport problems will be solved most satisfactorily.

Deficient rail services and coastal shipping has forced the steel industry to rely excessively on road transport. The highway system connecting the principal producing centers with the industrial steel-consuming areas has been reasonably efficient. Truck transport is used almost exclusively to deliver finished steel products to the southern, central, and northeast parts of the country. Apparently the higher cost of truck transport was more than offset by the direct delivery to the consumer and the inefficiencies of the other transport systems.

By the mid-sixties, the Brazilian government had organized a transport planning agency (GEIPOT) whose aim it was to study ways of modernizing the country's transport system. Various studies were under way to eliminate the bottlenecks in the railway and coastal shipping system in order to increase their effectiveness in serving such key industries as steel.

EFFICIENCY OF THE PRESENT LOCATIONAL PATTERN OF STEEL PLANTS

Tables 41 and 42 show two separate attempts to measure the locational advantage of various Brazilian firms, some market- and some resource-oriented. The evidence presented is not very conclusive. Obviously, the firms closer to the raw materials have lower assembly costs than the market-oriented firms, while the latter have lower delivery costs. The extremely low assembly costs of Belgo-Mineira and ACESITA in Table 41 are probably a result of the low cost of charcoal, while the low assembly cost of Volta Redonda is explained by the fact that the firm operates its own coal mines. Considering Usiminas, Cosipa, and Mannesmann in Table 41, it is hard to conclude that either market- or raw material–oriented location is superior when rail services are used in connection with the estimated assembly costs. If trucks are used to São Paulo, the advantage would seem to lie with Mannesmann. In the case of the Rio de Janeiro market, the assembly costs combined with rail transport seem to give the edge to raw material–oriented Belgo-Mineira. The trouble with these calculations is that the latter firm does not produce the same products as Volta Redonda, and Mannesmann could afford somewhat higher assembly and transport costs, given its special types of steel products; these have a much higher value than many of the products of Belgo-Mineira and Volta Redonda.

My own estimates in Table 42 are not as complete as the BAHINT estimates. I have used mainly information for the year 1964, and I have only included the transport costs of the major raw materials. The extremely low truck transportation costs for Acesita are because these estimates are based on data furnished directly by the firms; these rates are probably low because of unpublished rebates. The other estimates were indirectly arrived at and thus are probably too high, since I had no way of knowing discounts given to companies by trucking firms. Although my estimates seem to give the edge to raw material–oriented firms, these have to be qualified. I based my calculations on transport costs in 1964, when railroad rates were too low (railroad rates had not been allowed to rise at the same rate as the inflation) and when throughout much of that year the import of petroleum was still subsidized. By 1965, these price distortions in transportation rates had already been eliminated.

The conclusion one reaches from the above exercise is that it is difficult to pass a definite judgment on the efficiency of the locational pattern of the industry. It is doubtful that the locational pattern is drastically inefficient. Changes in the rate structure of the various transport modes might alter the picture, but so might changes in markets. Obviously the raw material–oriented steel centers are in a good position

Table 41

Assembly and Product Distribution Cost of Selected Steel Firms: Bahint Estimates (October 1965)

(US$ per ton of finished products)

Firm	Destination a. Assembly Cost	São Paulo b. Railroad	São Paulo c. a. + b.	São Paulo d. Truck	São Paulo e. a. + b.	Rio de Janeiro f. Railroad	Rio de Janeiro g. a. + f.	Rio de Janeiro h. Truck	Rio de Janeiro i. a. + h.
CSN (Volta Redonda)	36.59[a]	5.20	41.79	5.50	42.09	3.00	39.59	3.50	40.00
Usiminas	40.55	14.00	54.55	18.60	59.15			12.50	53.05
Cosipa	51.00	3.00	54.00	3.20	54.28			10.00	61.00
Mannessmann	44.38	10.50	54.88	8.60	52.98	7.50	51.88	7.70	52.08
CSBM Monlevade[a]	22.15	10.50	32.65	8.60	30.75	7.50	29.65	7.70	29.85
Acesita	36.44	14.00	50.44	18.60	55.04			12.50	48.94

Source: Taken from Booz Allen & Hamilton International study for the World Bank and BNDE (unpublished).

[a] Low cost attributed to Volta Redonda by BAHINT is because the enterprise operates its own coal mines: this advantage over other firms was assumed to be US$ 3.54 a ton. Also a cost advantage was US$ 13.80 of scrap purchased on the outside. Belgo Mineira's (CSBM) cost is probably too low, since rates of final goods shipped were assumed to be the ones listed from Belo Horizonte on.

Table 42

Transport Cost of Raw Material Inputs and Product-Distribution Cost
for Selected Firms in 1964, Author's Estimates

(in US$ per ton of ingot)

Firm	Destination a. Raw materials transport costs	São Paulo b. Railroad	São Paulo c. a. + b.	São Paulo d. Truck	São Paulo e. a. + d.	Rio de Janeiro f. Railroad	Rio de Janeiro g. a. + f.	Rio de Janeiro h. Truck	Rio de Janeiro i. a. + h.
CSN	17.52	4.14	21.56	4.39	21.91	2.16	19.68	2.93	20.45
Usiminas	12.85	11.01	23.86	8.34	21.19	8.28	21.13	8.35	21.20
CSBM (Monlevade)	3.22	9.77	12.99	17.74	27.51	6.66	9.88	10.47	13.69
Acesita	6.05	11.01	17.06	8.34	14.39	8.28	14.33	8.35	14.40

Source: Estimated from direct-input information from firms, railroad rates obtained from RFFSA, and from trucking rates published in *Boletim de Custos*. Usiminas information is for 1965. Raw-material estimate in this table is not comparable to assembly cost of previous table because a number of items are not included, such as electric power costs. The raw-material-transport cost estimate was reached by taking principal raw-material inputs of each firm, measuring the distance they had to travel by either railroad or truck and multiplying the by estimates of rates.

to service growing export markets (Usiminas shipping its products abroad via Vitoria) and also to service a possibly growing market in Central Brazil.

It will be interesting to watch the development of the area between Belo Horizonte and Vitoria in the rest of the twentieth century. The former city, where Mannesmann is located, also has a large industrial city manufacturing diverse products. As one travels in the direction of Vitoria through the Rio Doce valley (see map at the end of this volume), one encounters a whole series of steel enterprises—Belgo-Mineira's small Sabara works, the iron tube producer Cia. Ferro Brasileiro, the Belgo-Mineira's Monlevade works, and, further on the Acesita plant, then Usiminas and in Vitoria the Ferro e Aço Vitória works, and, of course, the installations of the Cia. Vale do Rio Doce, both at the mines and in Vitoria, with the new giant iron export port of Tubarão and the company's new pelletizing plant. There can be no doubt that this area has the makings of a Brazilian "Ruhr."

THE FOREIGN EXCHANGE COST
AND BENEFITS OF PRODUCING STEEL IN BRAZIL

Given the fact that Brazil's relatively low production costs of steel could be even lower with more use of imported coal, that scale economies will make themselves felt as the domestic and foreign market expand, and that the industry's locational distribution is not drastically inefficient, the question of opportunities foregone may still linger in the reader's mind. I shall only partially answer this question by considering Brazil's input advantages relative to other countries and by trying to measure the returns foregone by using iron ore domestically instead of exporting it. I shall argue below that I find little use for engaging in an exercise which tries to consider the alternative uses to which all the capital invested in steel over the last forty years might have been put.

There can be little doubt that Brazil has some basic input advantages for the production of steel. Table 43a shows the very low cost of iron ore. The Brazilian price is the price at the pit, which does not differ much from the United States price at the mine. The quality of the Brazilian ore, however, is substantially superior to that available almost anywhere else in the world. Also, the Brazil coal input price could be substantially lower if a greater proportion of imported coal could be used. Table 43b is instructive, since it gives evidence of a much lower cost burden of the "basic inputs" (raw materials and labor) compared with Europe or the United States (other production costs include energy, maintenance, and internal transport.) These costs and the higher administrative sales costs may be assumed to fall in the future with greater organizational efficiency. The greater financial burden is also because Brazilian firms, more than American or European firms, were much more dependent on external finance for their investments.

Table 43

a. Basic Imput Prices for Steel Inputs (1963)

(current dollars per ton)

	Argentina	Brazil	Chile	Mexico	Peru	Venezuela
Iron ore	14.55	2.82	6.45	7.50	7.63	4.81
Coal	17.69	18.33	18.33	8.00	25.75	19.00

Source: Naciones Unidas, Comision Economica Para America Latina, *La Economia Siderúrgica de America Latina*, p. 193.
Notes: For Argentina the iron ore price is the imported price; for Brazil and Chile coal prices are based on weighted average price of imported and domestically used coal and coal for Peru is based on price of imported coke.

b. Cost-Distribution for Steel Products in Brazil, Europe, and the United States

(percentage distribution)

	Brazil	Europe	United States
Raw materials	31	44	37
Labor	10	18	35
Other production costs	22	17	14
Administrative and sales costs	10	7	5
Depreciation	7	5	5
Interest	11	4	1
Taxes (excluding income tax)	9	5	3
	100	100	100

Source: "Perspectivas Da Participação Brasileira No Mercado Internacional do Aço" by Paulo Dias Veloso, July 1967, unpublished paper; information based on data collected by Booz Allen & Hamilton while working for 27 steel companies in 10 different countries.

In Table 44 I have tried to make an estimate of the foreign exchange returns foregone by producing steel domestically. Part a. of the table contains an estimate of the foreign exchange earnings foregone by the domestic consumption of iron ore. Annual domestic consumption was valued at the average export price of the respective years. Since Brazil furnishes only about 5 percent of the world market with iron ore, it was assumed that if the quantity of steel domestically consumed had been exported, it would not have changed the world price. In part b. of the table I have made an estimate of the value of domestic steel production in world dollar prices. I did not use domestic production at average import prices (column B), since Brazil imports special steels of higher value, which cannot be domestically produced and whose average price is much higher than the average for the common mix of steel. I thus took the average value of steel exports of a number of industrial countries, even though these might be too low because of occasional dumping behavior (see column C). From this I subtracted twice the value of imported coal, since coal is the principal inported input used in steel and is known to average about half of the value of imported inputs. The resulting value for imported inputs is still a little high, since we assumed in column D that all coal imported is used for steelmaking. The last column F is an estimate of the net value added of the Brazilian steel industry. This should be compared with column F of part a. of the table, the foreign exchange earnings foregone by using iron ore domestically.

Table 44 shows that in 1965, for example, Brazil would have had to import US$ 315 million worth of steel products had it foregone steel production, and had it used all the principal tradable input (iron ore) for exports, the country would have gained US$ 43.9 million, resulting in a negative balance of US$ 271 million. The sheer dimensions here convince me that the opportunity cost of steel production is relatively small. This impression is not lessened when one takes into account that the foreign interest and amortization payments of the industry in 1965 amounted to approximately US$ 35 million. It might also be suggested that an additional foreign exchange burden to be considered is that part of the depreciation that must be imported. Taking into account that because of the development of the capital goods industry, an increasingly larger proportion of Brazil's investment in the steel industry is produced domestically, the foreign exchange burden of replacing depreciated machinery has been estimated to attain at most 45 million dollars a year.

FOREIGN EXCHANGE BENEFITS VERSUS DOMESTIC
RESOURCE COST OF STEEL PRODUCTION.[7]

Table 45 below presents a different analysis of the costs of steel

7. This approach was suggested to me by Joel Bergsman.

Table 44

a. Value of Domestic Iron Ore Consumption in Brazil

	a Total iron ore mined (1000 tons)	*b* Total iron ore exports (1000 tons)	*c* Value of iron ore exports (in 1000 US$)	*d* Iron ore export price (C:B)	*e* Domestic iron ore consumption (A–B) (1000 tons)	*f* Value of domestic iron ore consumption at export prices (ExD)
1963	11,218	8,207	70,417	8.58	3,011	25,834,380
1964	16,962	9,730	80,638	8.28	7,232	59,880,960
1965	18,160	12,731	102,979	8.09	5,429	43,920,610

Source: Calculated from *Anuário Estatístico do Brasil*, 1966.

b. Value of Brazilian Steel Products Net of Imported Inputs

	a Output of rolled products (in tons)	*b* Rolled products valued at average import price (US$)	*c* Output of rolled products valued at average world price (US$)	*d* Value of total Brazilian coal imports (US$)	*e* Total imported inputs (2 × D)	*f* Net national value added of Brazilian steel (C–E)
1963	2,163,091	553,751,296	324,463,650	13,606,000	27,212,000	297,251,650
1964	2,414,797	545,744,122	362,219,550	24,376,000	48,752,000	313,467,550
1965	2,344,895	527,601,375	351,734,250	18,329,000	36,658,000	315,076,250

Source: Calculated from data in *Boletim, IBS*, Março 1967; United Nations, *Commodity Trade Statistics; Anuário Estatístico do Brasil,* 1966.

production at Volta Redonda in 1963. In this calculation, costs are taken from the national rather than the private viewpoint. That is, the cost of iron ore is taken at the export price, rather than at the lower domestic price. The cost of coke is taken as though only imported coal were used, because the higher cost incurred by using domestic coal is a subsidy to the domestic coal industry from the nation's point of view. The value of the product was estimated by taking the average

Table 45

National Costs of Volta Redonda's Steel Production in 1963

	Cost per ton of final product	
Item	Foreign exchange (US$ per ton)	Local costs (Cr$ per ton)
Blast furance		
Iron ore[a]	US$ 7.80	
Sinter[b]	5.85	
Coke[e]	16.00	
Other raw materials		Cr$ −788
Direct labor		573
Indirect costs		2,380
Gross capital charge[d]	7.55	2,200
Steel shop		
Additional raw material (mostly scrap)		12,342
Direct labor		1,514
Indirect costs		12,900
Gross capital charge	7.40	2,160
Rolling mill		
Additional raw materials (mostly scrap)		−6,827
Direct labor		2,019
Indirect costs		7,150
Gross captial charge	29.50	8,660
Administration and sales[e]		11,000
Total cost of sales	US$ 74.10	Cr$ 55,000

a. Iron Ore: physical requirement from CEPAL. Price is average export price, FOB Brazil, 1963 (US$ 8.58/ton).

b. Sinter: physical requirement from CEPAL. Since iron ore represents something less than half of the cost of sinter, the price was raised by half as much as was the price of iron ore.

c. Coke: The CEPAL cost per unit of product was reduced by a factor of .685. This reflects a 25 percent reduction in the coke rate and a 17 percent reduction in the price of coal. It implies that domestic coal costs 75 percent more than the imported coal per ton of pig iron produced. Changes in blast furnace capacity are not considered here.

d. Gross capital charge for blast furnace: CEPAL estimates were reduced by 32 percent, to remove the effect of using 40 percent of domestic coal. The total effect of adjustments for using 40 percent of domestic coal is to reduce the cost of pig iron by US$ 9.50 per ton, or 20 percent. Gross capital charges, in general: The CEPAL estimates were used; they are calculated as 15 percent of total value of investment as estimated by CEPAL. Their estimate is only a very rough guess; unfortunately it was not possible to obtain better estimates. The latter imply a capital-production ratio of more than US$ 500 per ton of rolled products. In value terms, this is more than 3 to 1. Since the blast furnace and the steel shop of Volta Redonda were operating fairly near effective capacity, this estimate does not seem likely to be too low; it must be admitted, however, that this is certainly the weakest part of the data. It should be mentioned that in its expansion plans announced in the mid-sixties, Volta Redonda's management estimated that capacity increase from 1.4 to 3.5 million tons would imply only an investment cost of US$ 232 per ton, and that a new 3.5 million plant would cost US$ 430, per ton. (see Companhia Siderúrgica Nacional, *Expansão De Volta Redonda, Plano D*, Rio de Janeiro, 1965, p. 9). Total capital charge was divided into 70 percent foreign exchange costs and 30 percent domestic costs, which is the approximate breakdown suggested by CSN personnel for past investment.

e. Administration and sales costs were estimated at approximately 10 percent of total cost of sales. This figure is based on rough estimates from various Brazilian firms, including CSN. All other costs are taken directly from the CEPAL estimates. Direct inquiry in 1967 at Volta Redonda yielded information on all costs except capital charges. The data were quite close to the CEPAL figures.

prices for the products, produced by Volta Redonda, in Japan, Germany and the United States. The analysis is in the form of calculating the exchange rate at which Volta Redonda is transforming cruzeiros (local costs) into dollars (net foreign exchange saved). The basic data are from the CEPAL study *La Economia Siderúrgica de America Latina* (mimeographed February 1966); these data were adjusted as explained in the notes to the table.

The value of the product, US$ 156/ton was estimated from the 1963 CSN sales mixture and rough averages of 1967 prices in Germany, Japan, and the United States. These prices were perhaps 10 percent lower in 1967 than in 1963, so here the estimate is quite conservative. The basic data are as follows:

Item	% CSN Sales, 1963	1967 price, Germany, Japan, United States
Rails	2.1%	US$ 130
Other nonflat shapes	7.7	140
Heavy plates	13.1	135
Hot rolled sheets	25.5	145
Cold rolled sheets	25.8	160
Galvanized	4.6	200
Tin plate	16.8	180
	95.6%	
Average price per ton		US$ 156

The value of the product, as explained in the notes to Table 45, was conservatively estimated at US$ 156. These calculations imply that the net cruzeiro cost was Cr$ 55,000 per ton, and the net foreign exchange saving was US$ 82 per ton. Thus the equivalent exchange rate for Volta Redonda steel was Cr$ 670 per dollar. This can be compared with the average import rate of 768 and the average export rate (excluding coffee) of 553. It seems therefore that the operation of Volta Redonda was efficient in a comparative cost sense, or, better, would have been efficient if the requirement for using 40 percent domestic coal had not existed.

This conclusion should be qualified in many ways. On the positive side, we should remember that Volta Redonda was the first large integrated coke-based steel mill built in Brazil. Furthermore, by 1963 it was already 17 years old, and its technology was already partly obsolete (especially in the steel shop). We had expected its efficiency to be lower, and would have been prepared to justify some inefficiency on grounds of learning effects. It is striking that this was not necessary.

On the other hand, various negative factors were not included in the analysis. Operation in early years may have been considerably less efficient than in 1963. We have no data on this. No charges were made for special infrastructure costs, such as workers' housing and local roads, as described above. We believe such costs should be charged to a "general costs of development" account rather than to the steel industry; others may disagree.

It appears that the two newest mills, Cosipa and Usiminas, will be more efficient than Volta Redonda when they reach the capacity of one million tons per year. This will happen in the early seventies. Their actual capital costs were each about 325 million dollars for about 650,000 tons of ingot capacity (and 1.5 million tons in the rolling mills) and plans for expansion to one million tons project additional investments of less than US$ 70 million. Thus, their capital charges should be about 20 percent lower than the CEPAL estimates for Volta Redonda.

Based on this evidence, it seems reasonably clear that Brazil's steel industry is competitive in the comparative cost sense, without any recourse to external benefits or—at this time—infant industry arguments.

It could be argued that had all the capital invested in steel gone into other sectors the rate of growth of the economy and/or of foreign exchange earnings might have been greater than it actually was. Investment resources could have gone into a new agricultural export sector, such as meat, or a lighter industry, such as textile or food products. Attempts to calculate what might have happened if resources used in steel had been employed differently involve the making of so many assumptions about markets, prices, and technology that such an exercise has very little meaning. With existing import restrictions in the developed countries against many light manufactured products, especially from Latin America, a substantial expansion of the light manufactured products capacity would not have made much sense. It is also most doubtful that the development of the steel industry took away resources that might have been used in developing the agricultural export sector or iron ore exports.

8

GROWTH PROSPECTS AND EXPANSION
PLANS OF THE BRAZILIAN STEEL INDUSTRY

THROUGHOUT THE first half of the sixties, a number of independent steel-demand projections were made in order to ascertain the capacity that would have to be created by the middle of the 1970s if Brazil is to be relatively self-sufficient in steel production. To examine some of these projections, the reader should recall some of the data on per capita steel consumption in Brazil and various other countries presented in Table 23. It was shown, for example, that Brazil's per capita ingot steel consumption was 44.2 kgs. in 1964, which represented about 40 percent of Argentina's per capita steel consumption, although Brazil's per capita income was about half of Argentina's. This comparison alone would lead us to expect a very high potential growth rate in steel demand. One would be led to suspect that the income elasticity of demand for steel is greater than one. Thus, with Brazil's population growth rate of more than 3.1 percent a year and a possible 3 to 4 percent a year growth rate of per capita income, much of it based on further industrial expansion, one would intuitively expect a vast expansion of steel demand, without even considering the country's future as a steel-exporting nation.

In 1964, M. Falcão made a projection of steel demand based on observations in the period 1947–63.[1] Per capita steel consumption was made a function of real per capita gross domestic product in constant 1960 dollars in the following way:

$$y = bx^a$$

where "y" is per capita ingot steel consumption, "x" per capita gross

---

1. *A Economia Siderúrgica Da América Latina: Monografia do Brasil.*

domestic product (in 1960 dollars) and "a" the income elasticity of demand for steel. The resulting equation obtained was:

$$\log y = 2.04581 \quad \log x - 3.53677 \quad r = 0.967$$
$$(.0646)$$

These results were used to make projections to 1975 and the values obtained are reproduced in Table 46a. Three alternative values were taken for the yearly per capita GDP growth rate, 2, 3 and 4 percent a year. Thus, the minimum demand in 1975 would be 9.9 million tons and the maximum 11.3 million tons. Of course, this leaves out of account any potential growth in the export of steel products.

Table 46

Some Alternative Projections of Brazil's Demand for Steel

a. *Falcão-CEPAL projections (made in 1964 based on 1947–63 data)*

	Per capita steel consumption (kg.)[a]	Steel demand (in 1,000 tons of ingots)		
		a	b	c
1965	53.90	4,665	4,382	4,099
1966	57.47	5,120	4,809	4,498
1970	72.97	7,293	6,841	6,399
1975	98.55	11,317	10,630	9,943

Source: CEPAL-Falcão Brazilian steel monograph, *op. cit.*, ch. 2.

a—assuming a per capita GDP growth rate of 4 percent a year.
b—assuming a per capita GDP growth rate of 3 percent a year.
c—assuming a per capita GDP growth rate of 2 percent a year.

a. Per capita projection based on assumption of GDP per capita yearly growth rate of 3 percent.

b. *Other Steel Consumption Projections*

(millions of tons)

	1965	1969	1970	1975
BNDE (1960)[a]	4.3	6.1		
Consultec (1961)[b]	4.2		6.2	9.2
S. P. L. (1962)[e]	5.2		10.2	
Teconometal (1963)[d]	4.9		7.9	12.7
BNDE (1965)	4.1	5.7	6.3	9.7
	(3.3)	(4.7)	(5.1)	(7.9)

c. *Falcão-CEPAL Projections for Selected Products*
(in 1,000 tons of ingots)

Table 46 (Continued)

	1965		1970		1975	
Nonflat Products	2,322	(53.0)	3,489	(51.0)	5,315	(50.0)
Flats	2,060	(47.0)	3,352	(49.0)	5,315	(50.0)
Of which tin plate	(351)	(8.0)	(547)	(8.0)	(850)	(8.0)

Source: Same as a. Same methodology was applied to individual products as was done for global projections.

Source: Same as part a. and BNDE, Departamento Econômico, *Mercado Brasileiro de, Aço,* Rio de Janeiro: Junho 1965.

Note: Dates next to institutions represent year the projection was made.

　　a. Obtained through exponential function, based on data between 1947 and 1959 $Y = 1,091.455 \times 1.092^x$, where Y is apparent consumption, and x is time.

　　b. $C = 0.003\ 227\ R^{1.0861}g^{1.2088}$ where c is per capita steel consumption, R index of real per capita income and g represents degree of industrialization.

　　c. Based on a study of consumption by sectors.

　　d. Based on two exponential equations, one for flat products $Y = 524.8 \times 1.1146^x$ and one for nonflat products (excluding rails) $Y = 657.5 \times 1.0875^x$ where Y is apparent consumption and x is time (1956 being the initial year)

　　e. Based on linear regression between steel consumption and a real product made up by averaging an index of manufacturing output and construction for data from 1947 to 1961 $Y = 14.6 + 0.9926\ X$, where Y is steel consumption and X is an index averaging manufacturing output and construction. The numbers in parentheses represent a lower projection level. It is based on the yearly growth rates obtained from the projection which are applied to the lower 1965 consumption level of 3.3.

Table 47

BAHINT Projections

a. *BAHINT Projection by Product* (1,000 tons)

	1966	1970	1975
Flat Products	1,269	1,795	2,692
(tin plate)	(227)	(308)	(446)
Nonflat products	1,383	2,006	2,899
Semifinished products	20	27	42
Total	2,672	3,828	5,633
Ingots necessary for total output	3,498	5,010	7,377

Source: Booz Allen & Hamilton International projections as published in Plano Decenal de Desenvolvimento Econômico e Social, Tomo V, *Indústria e Mineração, Serviços,* volumes 3, 8, 9 e 10, Rio de Janeiro, Ministério Do Planejamento e Coordenação Econômica, Março 1967.

b. *Global Steel Consumption Projections* (1,000 tons)
　　Based on following growth

rates per year	1966	1970	1975
9.7　(1956/8–1962/4)	3,538	5,123	8,139

Table 47 (Continued)

8.7 (1945/7–1962/4)	3,538	4,940	7,497
7.6 (1945/7–1964/6)	3,538	4,742	6,839
10.0 (BAHINT optimistic)	3,538	5,204	8,361
8.65 (BAHINT most probable)	3,538	5,069	7,459
8.0 (BAHINT conservative)	3,538	4,813	7,072

Source: Same as a; first three rates are based on average in period indicated in parentheses.

c. BAHINT Sectoral Projections of the Steel Market-Apparent Consumption
(1,000 tons)

	1965		1975	
Automobile	272	(12.6)	704	(12.5)
Railroad	161	(7.4)	256	(4.6)
Shipbuliding	44	(2.0)	127	(2.2)
Highway equipment	14	(0.6)	34	(0.6)
Agricultural equipment	28	(1.3)	77	(1.3)
Canning	189	(8.7)	446	(7.9)
Container	84	(3.9)	226	(4.0)
Domestic appliances	79	(3.6)	202	(3.6)
Commercial equipment	29	(1.3)	63	(1.1)
Construction	565	(26.1)	1,629	(28.9)
Industrial machinery	145	(6.9)	307	(5.5)
Fabricating equipment	176	(8.1)	428	(7.6)
Wire Drawing	300	(13.8)	960	(17.1)
Miscellaneous	80	(3.7)	174	(3.1)
Total	2,166	(100.0)	5,633	(100.0)

Source: Siderúrgia-Metalurgia-Mineração, São Paulo: Editora Banas, S.A., Junho de 1967, p. 124.

Table 48

Steel-Demand Projection Based On International Cross-Section Study

	1961	1965	1970	1975
Population (millions)	73.1	81.0	93.3	107.2
GDP in US$ millions	20.2	23.5	31.5	42.1
GDP per capita	277	290	337	392
Consumption of ingot steel per capita (tons)	46.5	49.1	58.7	70.3
Total consumption of steel (1,000 tons)	3,398	3,979	5,476	7,534

Source: Same as Table 45.

Note: Assumed growth rate of GDP between 1964 and 1975 is 6 percent.
Also per capita ingot consumption $= K$ $\underline{GDP}$ a, where $K = 1,2455$ and a $= 1,1926$.

Table 49

Demand Projections Used as a Basis for Brazil's Steel-Investment Program

(millions of ingot tons)

	1966	1970	1975	1976
Flat products	1,733	2,485	4,104	4,514
(tin plate)	(311)	(432)	(686)	(755)
Nonflat products	1,788	2,631	4,150	4,565
Total	3,501	5,126	8,254	9,079

Source: Same as Table 45.

Part b. of Table 46 shows other independently made steel-demand projections carried out in the period 1960–65. It will be seen that most of these projections also fall in the range of the Falcão-CEPAL projections. Special note should be taken of the numbers in parenthesis of the 1965 BNDE projections. These are based on the relationships obtained in the period 1947–61, but applied to the consumption base of 1965, when Brazil was experiencing an industrial recession. The resulting demand projection for 1975 is thus much lower.

Table 46c shows the Falcão-CEPAL product projections. These are especially interesting compared with the projections made by the firm of Booz-Allen & Hamilton International (BAHINT) shown in Table 46. The product projections of BAHINT are based on a survey made in 1965 and early 1966. Based on past trends, but also on the situation in 1965–66, when steel consumers in various sectors were interviewed, a projection was made of steel demand in various steel-consuming sectors (see Table 47c). It will be noted that no drastic changes in the structure of the distribution of steel-consuming sectors was predicted by BAHINT. Let us examine how BAHINT arrived at that conclusion by examining its procedure in making predictions for a number of individual sectors.[2]

The automobile sector was not expected to absorb a larger proportion of steel output in 1975 than in 1965 because it was found by BAHINT that the rate of growth of demand for automobiles was not expected to be faster than the rate of growth of over-all steel demand. The rate of growth of demand for automobiles was not assumed to be substan-

2. Detailed information on BAHINT's projections by sectors was obtained from *Siderúrgica, Metalurgia, Mineração.* 1967 (São Paulo: Editora Banas, S. A., Junho de 1967), pp. 122–123.

tially affected by the further development of highway construction and by the lower costs resulting from scale economies that were expected to manifest themselves in the late sixties and throughout the seventies. It was assumed that participation of the shipbuilding industry would not grow. At this writing (1968), this seems a doubtful assumption, since the Brazilian government in September 1967 declared that it meant to reduce substantially the country's overwhelming dependence on foreign shipping for its exports and imports. In that month, the government ordered 25 ships from the Brazilian shipping industry to expand the country's merchant marine.

Scarcely any change was predicted for the participation of the road-building equipment industry, which seems doubtful in view of the country's great need to improve and expand its highway system. The assumption that the agricultural equipment industry would not increase its participation is particularly surprising in view of the acute need in Brazil to modernize agriculture. Also surprising is the predicted decline in the share of the canning industry. Considering the surge in the urban population of Brazil which up to the mid-sixties had not been accompanied by widespread modernization of the food industry, one would expect a far greater participation of that industry.

Assuming a far larger growth of such industries as automobiles, shipbuilding, canning, and agricultural equipment, one would expect a higher growth rate of demand for steel consumption than is predicted in Table 47, and especially a much higher rate of growth of such sectors as tin plate. The global steel consumption projections in Table 47b apply alternative growth rates to the assumed 1966 ingot consumption base of 3.5 million tons.[3]

Table 48 contains a projection based on a world cross-section analysis which takes per capita income as the main independent variable. The results are quite similar to the BAHINT projection. A number of objections to this type of projection are, however, in order. For example, how relevant is it for a country of continental proportions, whose industrial production is located in the center-south of the country and whose main markets are in the same area, to take the over-all per capita income instead of the per capita income of the main steel-consuming areas?

Table 49 contains the demand projections which the Brazilian govern-

3. The actual steel production for 1966 was 3.76 million tons in terms of ingots, and the apparent consumption was about 3.95 million tons of ingot equivalents.

ment finally adopted in order to plan its investment program. It will be noted that the projection for the demand of final steel products in 1976 is 9.1 million tons, which implies an ingot steel capacity of about 11 million tons—not far from the original projections made in the early sixties.

A great weakness of all these projections is that they disregard the country's export possibilities. The experience of the first half of the sixties shows that it was possible for Brazil to export a significant proportion of its steel output.

Table 50 shows a substantial growth of exports in the mid-sixties. It is true that the great quantity of steel exported in 1964 and especially 1965 was a result of the industrial recession Brazil experienced in those years, which forced some firms to sell below cost to countries abroad. Most of the companies that exported (Usiminas, CSN, Cosipa, and Belgo-Mineira) in the mid-sixties, however, expressed hope of continued access to such markets as Argentina and the United States. Besides the usual obstacle of foreign protection which future growth of exports faces, another problem is the excess world steel-production capacity. In the mid-sixties, this led a number of key steel-producing countries to dump their products on the world market.[4] Given the good position of Brazil for producing steel at a relatively low cost, however, and given the possibilities of establishing an adequate export credit mechanism for financing steel exports and the possibility of establishing direct complementary relationships with foreign steel producers (to be discussed in the following section), there seems to be no reason why Brazil should not be able to export between 10 and 15 percent of its steel output by the mid-seventies.

Unfortunately, the export possibilities to other Latin American countries do not appear very promising. Each of the major Latin American

4. One good example of steel dumping operations is the case of Germany in 1966 (units US$ per ton):

	Internal Price	*Export Price (FOB)*
Plates	126	84
Hot rolled sheet	145	100
Cold rolled sheet	166	104

Similar information was available for Japanese products in early 1967:

Hot rolled sheet	147	95
Cold rolled sheet	158	115

Source: Annual Report of the European Coal and Steel Community, 1966; Japan Metal Bulletin.

countries that have undergone industrialization is building up a steel industry. Although the steel-export spurt in 1965 was a result in part of a substantial rise of exports to Argentina, the long-run outlook does not seem very favorable. Brazil exported semifinished steel products to that country because its rolling mill capacity was larger than the capacity of its blast furnaces and steel shops. Argentina intends to expand its pig iron and steel making facilities, however, in order to be more self-sufficient in the future. Unfortunately, the striving towards self-sufficiency in steel production is apparent also in such countries as Chile, Mexico, Peru, Colombia, and Venezuela.

Table 50

Brazilian Steel Exports

(tons)

a. *Total exports*

1962	10,513
1963	53,679
1964	249,830
1965	482,353
1966	149,618
1967 (1st Q)	98,000

b. *Geographical Distribution of Usiminas Exports in 1966—Total exports 87,450 tons*

Argentina	45%
United States	52
Uruguay	3
	100%

Source: Veloso, Paulo Dias, "Perspectivas da Participação Brasileira No Mercado Internacional do Aço," unpublished, Julho 1967.

BRAZIL'S STEEL-EXPANSION PLANS INTO THE 1970S

By the middle of the 1960s, a large number of projects had been developed both to expand existing facilities and to create entirely new steel plants. Since it was obvious by the mid-sixties that a substantial expansion of steel capacity was needed to meet even the most pessimistic levels of demand projected for the middle of the seventies, and since both public and private firms would need substantial official domestic and international financing, the Brazilian government and the IBRD contracted the consulting firm Booz Allen & Hamilton International to examine the industry and its expansion plans and to make recommendations as to what type of expansion was needed and what projects would be worth supporting. Table 51 reproduces the projects pending in 1966 and the projects recommended by BAHINT.

It will be noted that Volta Redonda had planned to expand to a level of 3.5 million ingot tons by 1975, while Cosipa and Usiminas had plans to reach the 2 million–ton level by that time. Belgo-Mineira intended to reach a level of 700 thousand tons, and Ferro e Aço de Vitória had plans to become an integrated plant with a capacity of at least one million tons. It can be seen that a number of large new projects had been contemplated. Açominas was a plan by the government of Minas Gerais to set up a newly integrated plant, and Cosigua was a project by the government of Guanabara to build a new tidewater plant near Rio de Janeiro. The Caemi plan was to set up a tidewater plant of 2 million tons near Vitória, which would export semifinished steel products to the rolling mills of a European or American partner in the operation. At this writing, this project would possibly be picked up by the government iron ore exporting firm Cia. Vale do Rio Doce, which would build such a plant at its port of Tubarão, near Vitoria. Again, this would be done in partnership with a foreign firm or group of firms, thus assuring a market. Should such a plan become a reality, it would mark an interesting new trend in the international division of labor along vertical lines.

Booz Allen's recommendations, as can be observed in Table 51, are considerably more modest. Except for two small firms, Usinor in Recife and Cosima in Mato Grosso, it excludes all new projects, and expansion plans for Volta Redonda were trimmed down to 2.5 million tons. The original plan of Volta Redonda included increased coke-production facilities, a new blast furnace, a new LD steel shop, a universal slabbing mill, a new hot-strip mill and associated equipment for cold rolling and finishing, a continuous galvanizing line. Booz Allen's recommendations eliminated the hot-strip mill and recommended modifications in the existing mill. It also eliminated a universal slabbing mill and recommended a lower-cost blooming mill. It has approved a new LD shop, but not the oxygen plant for the new steel shop.[5] Cosipa's and Usiminas's plans were also cut to a level of one million tons, large enough for these enterprises to reach adequate scale economies to make them profitable.

5. One steel expert has questioned the value of expanding Volta Redonda at its present location instead of CSN building an entirely new plant at a different location.

Table 51

Alternative Expansion Projects of the Brazilian Steel
Industry for 1975

(1000 tons of ingot ton capacity)

	1966 Capacity	1975 Planned capacity	BAHINT recommendations
Volta Redonda (CSN)	1,400	3,500	2,500
Cosipa	625	2,000	1,000
Usiminas	634	2,000	1,000
Belgo-Mineira	400	700	520
Ferro e Aço de Vitória	70[a]	1,000	300[a]
Acesita	120		222
Riograndense	160		200
Barra Mansa	90		160
Lanari	30		90
N. S. Aparecida	28		64
Cosima	0	140	50
Usinor	0		120
Caemi	0	2,000	
Açominas	0	1,000	
Cosigua	0	1,000	
Usiba	0	180	
Anhanguera	70	200	70
Mannesmann	220	435	220
Villares	70	80	70
Aliperti	180	540	180
Minibrasil[b]	180	400	180
Other Projects	0	381	
	4,277	15,556	6,946

Source: Same as Table 43.

a. Both actual and BAHINT recommended capacity refers only to rolling capacity of semifinished products from other mills. FAV's own expansion plans refer to an integrated mill of one million–ton ingot capacity.

b. These are a group of small plants belonging to the firm Mineração Geral do Brasil which have been idle since the firm went into bankruptcy in 1964.

Note: 1975 Planned Capacity–capacity levels which individual firms plan to reach by 1975. BAHINT Recommendations—total capacity levels which BAHINT recommends by 1975.

Belgo-Mineira's expansion plans were also trimmed from 700 thousand to 520 thousand tons, and BAHINT did not recommend the expansion of Ferro e Aço de Vitória into an integrated mill producing nonflat products (mainly medium and light sections), but only recommended the expansion of its rolling facilities.

The general picture that emerges from the BAHINT recommendations is that the expansion of the steel industry into the seventies should consist largely of the expansion of existing facilities rather than the establishment of any new major enterprises. Also, most of the addition to facilities recommended (with the exception of Volta Redonda) consisted of a rounding-out of existing capacity rather than the expansion of major firms into new lines of production.

The estimated cost of the BAHINT program is shown in Table 52. Since this investment program would increase the country's ingot capacity by about 2.4 million tons and thus the output of final goods by about 2.2 million tons, the latter valued at about US$ 120 to 130 dollars per ton would give an increment in output valued almost at US$ 300 million. This means that the BAHINT program would imply a capital/output ratio of about 2. This would represent a relatively inexpensive expansion program, since, as already mentioned, it would amount basically to an expansion of existing facilities. The capital-output ratio is especially high when new installations have to be constructed because of the need for a large quantity of civil construction and basic infrastructure. This is not necessary in the type of program envisioned by BAHINT.

It seems that the BAHINT recommendations will not be adhered to, nor will all the original expansion projects and new projects. Certain projects not considered by BAHINT as viable for the expansion of the steel industry will probably go through under political pressures—the steel mill planned in Bahia (Usiba), for example—or because government circles responsible for obtaining finance for the steel-expansion program will accept more optimistic forecasts about the demand for steel than the one assumed by BAHINT in planning the expansion program. Also, the fact that the IBRD decided not to finance the steel expansion program (the motives for this are not clear, although the excess world steel capacity in the mid-sixties seems to have been an influential factor in the decision) will make the BAHINT program less binding. If the Cia. Vale do Rio Doce finds a willing foreign partner, there seems a good chance that a new 2 million–ton mill in Vitoria will go up for the exportation of semifinished products. Expansion of exports and a new spurt of industrial production in the late sixties might also convince authorities to expand Volta Redonda, Usiminas, and Cosipa beyond the limits set by BAHINT.

One thing seems clear, however, that by the seventies Brazil's steel production will be large enough to make full use of economies of scale.

Table 52

Investment Cost of BAHINT Expansion Plan

(in US$ millions)

Firm	Additional Capacity (1,000 tons)	Equipment	Total Cost
CSN	1,100	243.4	294.0
Cosipa	375	50.6	62.1
Usiminas	366	53.5	69.7
Belgo-Mineira	120	19.0	25.2
Acesita	102	29.2	40.7
Lanari	60	3.6	5.8
Aparecida	36	4.2	8.1
Barra Mansa	70	2.0	4.0
Riograndense	40	2.3	2.9
Usinor	120	40.7	49.6
Cosima	50	14.6	18.4
FAV		16.0	22.2
Total	2,439	479.1	602.5

Source: Plano Decenal De Desenvolvimento Econômico e Social, Ministério do Planejamento e Coordenação Econômica, Tomo V, *Indústria e Mineração, Serviços,* volumes 3, 8, 9 e 10, Rio de Janerio, Março 1967, p. 22.

This fact, combined with lower production costs with a lowering of the coke rate by the use of more imported coal and greater efficiencies in blast furnace and steel-shop operations, should make Brazil's steel industry competitive in world markets.

9

GENERAL CONCLUSIONS

THE BRAZILIAN experience in establishing a steel industry should make it clear that the usual condescending cliché about the wastefulness of implanting a heavy industry in a developing country has no universality.[1] Given the necessary natural resources, a large market, and an already trained technical elite in matters of steel technology, it is entirely possible for a developing country to establish a steel industry with a comparative advantage.[2]

We have seen that the establishment of a steel industry was not a new project conceived to accompany the industrialization drive of the country in the post–World War II period, but that it has long historical roots. The main reason for iron and steel production not developing more rapidly before the 1930s was that the economic structure of the country was oriented towards agricultural production. This resulted in there being only a small domestic market for steel products, and steel production for export in a predominantly agricultural economy was an impossibility, given the substantial capital needs for large-scale iron and steel production and the needs for a complementary infrastructure in power and transportation.

1. "Nationalism tends to emphasize investment in visible symbols of development —large irrigation projects rather than individual wells, large new modern factories rather than improvements to old factories—in preference to less visible but frequently more socially profitable types of investment," Harry G. Johnson, *Economic Policies Toward Less Developed Countries,* p. 68.

2. I realize that in a strict sense I cannot speak about comparative advantage, since I compared in this volume only costs of steel production. One would need a whole series of other industry studies comparing costs and productivity, to talk about comparative advantage as such.

We have also seen that throughout the nineteenth century and in the early part of the twentieth century Brazil was, given her resources, conscious of her iron- and steel-making potential. Brazil's age of iron- and steel-making on a large scale was thus anticipated by the founding of institutions for the training of engineers specialized in the study of minerals and metallurgy. Once sustained industrialization occurred, steel production developed rapidly. The industrial spurt of the First World War awakened the consciousness of the country to its industrial future and the growth of a number of small steel concerns in São Paulo and Minas Gerais was closely related to the fact that industrialization was not necessarily a short-term emergency phenomenon to make up for wartime shortages. The entrance of foreign capital in the twenties to build Belgo-Mineira only occurred because of a belief in Brazil's industrial future.

The pressure throughout the thirties for the establishment of a large, integrated steel mill which finally resulted in the creation of Volta Redonda shows that those elements of Brazilian society that believed in the long-run future of Brazil as an industrial country thought of the steel industry as a natural element in a growing industrial complex. Considering the country's vast iron ore resources, this is quite understandable. The early establishment of a steel industry also helps to explain the early vertical integration of Brazil's industrial complex, that is, the creation of a whole contingent of heavier industries accompanying the development of consumer-goods industries.

Brazil's steel industry was built by both private and public enterprises. The former concentrated on smaller establishments, producing mainly nonflat products and special steels. Only foreign private firms (namely, Belgo-Mineira and Mannesmann) had the resources to build medium-sized, integrated mills. Private domestic resources were not adequate to build large, integrated mills, especially those producing flat products. It was also shown that the Brazilian government only reluctantly came into the picture when it became clear that private foreign capital would not finance the establishment of large, integrated mills. The steel industry's example thus provides a clear case of a situation wherein the natural resource endowment and the market (given the general industrialization process) made the creation of large-scale steel establishments a viable undertaking, but where the socioeconomic development of the country was still such that private institutions did not exist to assemble the needed capital and provide the organization necessary for such an undertaking. This explains the reliance on foreign private capital and government enterprises to develop the large, integrated steel firms.

It could be claimed that capacity to assemble private savings is also a necessary endowment to be considered and that if private capital markets and organizational developments are still underdeveloped, a country should direct its resources into channels wherein the private sector is capable of functioning. Such an argument will have little influence on a government whose aim it is to promote rapid economic growth. If the road to industrialization, and hence rapid general economic growth, involves the development of certain sectors wherein private enterprise is not capable of functioning, any practical planner will not hesitate to involve the government.

Steel technology was imported from abroad. Each of the large plants was built by specialized foreign construction firms: Volta Redonda was planned and its building supervised by the American firm Arthur McKee, Cosipa by the American firm Kaiser, Usiminas by a group of Japanese firms, Mannesmann by the parent company in Germany. The plants, however, were run by Brazilian manpower after a relatively short period of time. Also significant, a larger proportion of the plants to be erected in the late 1960s and the 1970s will be produced within Brazil. Brazil has already acquired the capacity to build large blast furnaces, many parts of the rolling mill, and many technical installations. In addition, most large firms in Brazil devote some resources to research and have already made several contributions to the making of various types of steel.

The tradition of training metallurgical engineers even before the establishment of large-scale steel enterprises was partially responsible for the easy absorption of steel technology. The adaptation of a largely untrained labor force proved relatively easy. It is true that the expenditures of firms for the training of its labor (especially the large educational and other social programs that had to be established) were much greater than was the case in similar enterprises in more developed countries. A good "developmentalist," however, would consider this a worthwhile cost.

The very nature of a steel mill results in the training of a great diversity of manpower. A whole variety of engineers, skilled workers, and administrators are needed to look after the coke ovens, blast furnaces, steel shops, rolling mills, repair shops, foundries, transport network, and electrical installations. Not only is there a need to develop men of many specialties, but the organization of each section of a steel mill, the synchronization of the operations of these sections, the logistics of supplies, the planning and expansion of infrastructure serving the mills, changes the flavor of the industrial organization within which men work. There exists a substantial amount of evidence that the manpower training the steel industry imposed on the country has benefitted many other industrial sectors. In Brazil, the flow has usually been from steel to other industries. Many engineers, industrial administrators, and many skilled workers in a large proportion of Brazil's industrial enterprises (both private and public) received their initial training in the large government and private steel mills.

Steel technology is such that there is little choice between the economist's traditional concern about labor versus capital-intensive production techniques. Given the cost of establishing a large-scale steel mill and the scarcity of capital, it pays to adopt the most modern technology. All this, as we have seen, does not imply that steel has only a small direct and indirect employment impact. The establishment of the latest techniques, the use of LD steel shops, for example, also gives a country certain "advantages of backwardness"—the advantages of a latecomer. Given Brazil's steel installations, its first-rate iron ore, and its potential blast furnace productivity performance with only imported coal, the country could be one of the most efficient steel producers in the world. It is well known that many European plants and some sections of American steel producers are operating with some fairly outdated technology. This suggests the possibility of some degree of a new division of labor in the world. The possibility of establishing a semifinished steel products mill of 2 million–ton capacity to supply products for final rolling mills in the United States or Europe (see the previous chapter) certainly points in this direction.

It was seen that the locational pattern of the industry was not seriously distorted. The industry was located partially near the natural resources (iron ore), partially near the market, and partially in between. There is, however, little doubt that a more favorable location would have been achieved had there been no political side conditions imposed on the policy makers.

It could still be argued that the investment of resources used in developing the Brazilian steel industry might possibly have had an even greater impact had they been invested in an alternative sector, say a new agricultural export sector, such as meat, or a lighter industry, such as textiles or food products. Attempts to calculate what might have happened had resources used in steel been employed differently imply the making of so many assumptions about markets, prices, and technology, that such an exercise would not be very meaningful. We did show, however, that the opportunity cost in terms of iron ore exports foregone, compared with the steel-import savings, was extremely low. Also, with existing import restrictions in the developed countries against many light manufactured products, especially from Latin America, a substantial expansion of the light manufactured products capacity would not have made much sense. It is also doubtful that the development of the steel industry took away resources which might have been used in developing the agricultural export sector.

Finally, it was shown that for a country with the natural advantages for steel production and a large internal market, the growth of the steel industry had a substantial growth impact through its high linkages with other industries. And, given the complexity of a steel operation, the planning associated with the building of its numerous sections, the infrastructure in transportation and power supply associated with it, and the training its many specialized shops give to the labor force, the development impact of steel through external economies has been extremely high, though, of course, difficult to calculate.

APPENDIX I
Brazilian Iron and Steel Enterprises

In this appendix are listed 30 large and small Brazilian steel firms. It is not an all-inclusive list. About a dozen very small steel enterprises have not been included. Also not included are the new firms that have not gotten beyond the project stage. The basic idea of this appendix is to give the reader a feeling for the type of private and public enterprise that has sprung up since the early part of the twentieth century. Employment figures mostly refer to the year 1964 and include all employees.

Aços Anhanguera S.A. Location, Mogi das Cruzes, São Paulo. Founded, 1962. Began to produce in May 1966. A privately owned firm controlled by the ICOMI group (Ind. e Com. Minérios S. A.), with participation by Bethlehem Steel Corp., SKF group. The latter two provide technical assistance. The firm produces special steels. It has electric steel furnaces using scrap and pig iron with capacity of 90,000 tons a year, and a blooming mill, billet- and bar-rolling facilities.

Aços Laminados Itaúna S.A. Location, Itaúna, Minas Gerais. Founded 1952. A privately owned firm. Has one SM steel furnace and a foundry. Ingot capacity is 22,500 tons a year. It produces ingots for sale and a small quantity of iron castings. Employs 155.

Aços Villares S.A. Location, São Caetano do Sul, São Paulo (on the outskirts of the city of São Paulo). Founded, 1939. A private firm controlled by the Villares family. A producer of special steels. Originally founded to produce castings for Elevadores Atlas, S.A. (now Indústrias Villares, S.A.) Capacity of about 70,000 tons a year of special steels produced from scrap and pig iron in electric furnaces. Produces castings, forgings, and rolled products and in mid-sixties facilities were modernized to produce rolling-mill cylinders. Employs 2,040.

Acesita (Aços Especiais Itabira) Location, Coronel Fabriciano, Minas Gerais. Established in 1944 as an integrated steel plant for the production of special steels—bars, light shapes, silicon sheet, etc. Established first by private group headed by Percival Farquhar, but later acquired and still owned by the Banco do Brasil. Has following capacities (per year): sinter, 144,flfl tons; Pig iron, 120,000 tons; steel ingots, 120,000 tons; flat products, 38,400 and nonflat products, 112,500 tons. Has both blast furnace based on charcoal and an electric reduction furnace (has own hydroelectric generating plant). In steel shop has a Bessemer converter and electric furnaces. Employs 4,879.

Açonorte S.A. Location, Recife, Pernambuco. Established, 1958. Privately owned. Has an electric furnace steel shop and rolling mills for production of small bars, rods and shapes. Began rolling semifinished products of Volta Redonda, but started producing steel ingots in 1963. Ingot capacity per annum is 30,000 tons. Employs 371 workers.

Barra do Piraí S.A. Location, Barra do Piraí, Rio de Janeiro state. Founded, 1952. Controlled by a Swiss firm. Works with Elevadores Schindler. Electric steel furnace with yearly capacity of 15,000 tons. Employs 450.

Companhia Brasileira de Usiminas Metalúrgicas. Formed in 1925 by the family firm Hime & Co. and still basically family owned. Firm consists of two plants, Barão de Cocais, located in Minas Gerais, with a pig iron capacity of 70,000 tons a year, Usina de Neves, located in São Gonçalo, Rio de Janeiro, has a non-flat-products rolling capacity of 50,000 tons. The output of the latter consists of small bars, nails, and similar products. Total employment about 1,300.

Cobrasma, S.A. Location, Osasco, near the City of São Paulo. Founded in 1944. Controlled by a private group. Steel production based on electric furnaces. Capacity of the firm for steel foundry products is 20,000 tons a year and for forged products is 25,000 tons a year. The firm produces railroad wheels, axels, etc.

Companhia Brasileira do Aço. Location, City of São Paulo. Founded in 1943. Owned by a private group. Has an SM furnace and a rolling mill for billets and small shapes. The plant sells about half of its ingots and billets to to a re-rolling plant associated with firm. Ingot capacity is 18,000 tons a year and rolling capacity 6,000 tons. The firm employs 343.

Companhia Ferro e Aço de Vitória. Location, Vitória, Espírito Santo. Founded in 1942. First established as a blast furnace plant, it is currently a major re-roller of semifinished blooms provided by Usiminas, Volta Redonda, Acesita, Mannesmann, into light bars and medium structurals. Has a rolling capacity of 150,000 tons. Plans to integrate backwards. Is owned by BNDE, with a tiny participation of the German firm Ferrostaal, A.G., which provided technical assistance. The firm employes 1,022.

Companhia Ferro Brasileiro S.A. Location, Caeté, Minas Gerais. Founded in 1931. Owned by French interests. Producer of centrifugated iron tubes and iron foundry products. Pig iron capacity about 80,000 tons a year. About 2,095 employed.

Companhia Siderúrgica Belgo-Mineira. Founded in 1921, in Minas Gerais.

Sabará Plant—started by a private Brazilian group and acquired by the Belgian-Luxemburg concern ARBED, still the principal owner. An integrated plant with pig iron capacity (based on charcoal) of 60,000 tons a year; an SM steel shop with 50,000 capacity and a nonflat rolling mill of 60,000 capacity. Employs 1,422.

Monlevade Plant—built in the thirties and opened in 1936. Current capacity is 360,000 tons of pig iron (charcoal based) a year, ingot capacity of 400,000 tons (produced both by SM and LD shops); flat rolling capacity of 120,000 and nonflat of 176,000 tons a year. Employs 4,937.

Belo Horizonte Wire Mill—installed in the industrial city of Belo Horizonte in 1961. Capacity of 120,000 tons; employs 1,090.

Companhia Siderúrgica Nacional. Location, Volta Redonda, Rio de Janeiro State. Founded 1941. Owned mainly by Brazil's national treasury. Pig iron capacity 770,550 tons a year; ingot capacity 1,350,000 tons a year; flat products capacity 810,667 and nonflats 241,646. Has large steel foundry. Owns iron ore and coal mines. Employs 14,012.

Companhia Metalúrgica Alberto Pecorari. Two plants located in the city of São Paulo. Privately owned. Founded in 1961. Ingot capacity 18,000 tons. One plant has electric steel melting shop and rolling mill that produces billets for other plant. Employs 113 workers. (In 1965 steel melt shop was closed and plants worked as rerollers).

Companhia Siderúrgica Mannesmann. Location, industrial city of Belo Horizonte. Founded in 1952. Controlled by German parent company. Produces seamless tubes, medium and heavy bars in special and carbon steels. Pig iron capacity, 250,000 tons a year (has both electric reduction furnace and one blast furnace); ingot capacity, 320,000 tons (both electric and LD furnaces); nonflat rolling capacity, 250,000 tons and seamless tube capacity of 124,000 tons. Employs 4,728.

Companhia Siderúrgica Pains. Location, Divinópolis, Minas Gerais. Founded in 1953. Owned by private group. Has one SM shop which receives liquid pig iron from a nearby blast furnace by truck. Its rolling mill produces light bars. Its ingot capacity is 30,000 tons a year. Employs 524.

Companhia Siderúrgica Paulista (COSIPA). Location, near Santos, São Paulo. Founded in 1953. Rolling facilities started to produce in December 1963 and only by early 1966 did all sections function. Owned by BNDE. Pig iron capacity, 600,000 tons a year; ingot steel (LD furnace) 750,000 tons. Flat rolling capacity about 1,500,000 tons. Employs 5,600.

Companhia Metalúrgica Barbará. Location, Barra Mansa, state of Rio de Janeiro. Founded in 1937. Owned by Barbará family and a group of French companies. Producer of centrifugated tubes, iron foundry products. Capacity for tube production 100,000; iron foundry capacity, 15,000. Employs 756.

M. Dedini S.A. Located in Piracicaba, São Paulo. Founded in 1920. Owned by family which has interests in firm building machinery for sugar industry, including motors, transformers. Produces iron and steel castings for other firms, also billets, bars, shapes, wire rods. Has two SM steel furnaces and one electric furnace. Ingot capacity is 120,000 tons a year and nonflat capacity is 70,000 tons; steel foundry, 6,000 tons and iron foundry, 10,000 tons. Currently building a charcoal blast furnace. Employs 824.

Lafersa M.A. (Laminação de Ferro S.A.) Location, industrial city of Belo Horizonte, Minas Gerais. Founded in 1953. Privately owned. Installations include one charcoal blast furnace and one SM steel shop. Rolling mill produces round bars and wire rods. Pig iron capacity, 25,000 tons; ingot steel, 22,000 tons. Employs 353.

Indústria Metalúrgica N.S. Aparecida S.A. Location, Sorocaba, São Paulo. Founded in 1944. Privately owned. Installations include an electric steel furnace, nonflat products rolling mill and forging installations. Ingot capacity, 28,000 tons a year; nonflat products capacity, 38,000 tons. Employs 478.

Mineração Geral do Brasil Ltda. Established in 1938 and under the control of the Jafet family. The family collected a series of plants started under other sponsorships. Since February 1965 the firm is in receivership and most plants have been shut down since. The founding date next to each plant refers to the original construction of each plant.

Usina de Mogi. Location, Mogi das Cruzes, São Paulo. Has two charcoal blast furnaces, five SM steel furnaces and rolling mill for light bars and seamless tubes. Pig iron capacity, 90,000 tons a year; ingot capacity, 180,000 tons; nonflat rolling capacity, 110,000 tons. Employed 2,180.

Usina São Caetano. Location, São Paulo. Founded in 1925 and taken over by Jafet group in 1950. Acquired from Jafet group in 1965 by SAAD do Brasil. Pig iron capacity, 50,000 tons a year; ingot capacity, 48,000 tons; nonflat capacity, 30,000 tons. Produces billets, wire rods. Employs 952.

Usina São Francisco. Location, São Caetano do Sul, São Paulo. Founded in 1944. Produces light bars and shapes from own ingots, produced in electric furnace. Ingot capacity, 27,000 tons; nonflat rolling capacity, 50,000. Acquired by Jafet group in 1959. Employed 420.

Usina Santa Olímpia. Location, Ipiranga, São Paulo. Founded in 1943 and acquired by Jafet group in 1959. Produces light bars and shapes from ingots produced in its electric furnaces. Ingot capacity, 40,000 tons. Employed 680.

Usina De Martino. Location, Ipiranga, São Paulo. Founded in 1944 and taken over by Jafet in 1959. Produces billets and tubes. Ingot capacity, 30,000 tons; nonflat capacity, 20,000 tons. Employed 446.

Usina São José. Location, Santo André, São Paulo. Founded in 1943 and acquired by Jafet in 1959. Produces ferro alloys, light bars and shapes. Ingot capacity, 40,000 tons. Nonflat capacity, 80,000 tons. Employed 543.

Usina Nova Iguaçu. Location, Nova Iguaçu, State of Rio de Janeiro. Founded 1938. Produces Ferro alloys. Employed 161.

Companhia Siderúrgica Honório Gurgel. Location, Guanabara. Founded, 1938. Produces ferro alloys. Employs 181.

Lanari S.A. (Indústria e Comércio) Location, Paracambi, State of Rio de Janeiro. Founded, 1945. Family owned. Started as a steel sales company. In 1952 started rerolling operations for concrete-rod products with billets furnished by Volta Redonda. In 1958 a cold-charge SM steel shop was added with ingot capacity of 30,000 tons a year. Employs 767.

S.A. Comércio e Indústria Souza Noscheze. Located in the City of São Paulo. Founded in 1920. Family owned. Has two electric furnaces which provide castings and ingots which are either sold to others or rolled into light bars. Capacity, 30,000 ingot tons a year. Employs about 800.

Siderúrgica Barra Mansa S.A. Location, Barra Mansa, State of Rio de Janeiro. Founded, 1937. Owned by private groups. Pig-iron capacity, 90,000 tons; ingot capacity, 90,000 and nonflat rolling capacity, 140,000 tons.

Usina Queiroz Júnior S.A. Has two plants. One located in Itabirito, Minas Gerais (Usina Esperança) and the other fairly close by (Usina Gagé). Founded in 1891. Esperança produces iron for its foundry and also has a steel foundry operation. The Gagé plant produces pig iron. Pig iron capacity of Esperança is 47,000 tons and Gagé is 28,000. Iron foundry has capacity of 6,000 tons and steel foundry of 6,000. Esperança employs 1,154 and Gagé 291.

Usiminas. Location, Ipatinga, Minas Gerais. Founded in 1956. Principal share owner is BNDE. Japanese group is large minority owner (owning about 20 percent of shares). Began operations in 1962. Pig iron capacity, 600,000 tons; ingot capacity (LD furnace), 624,000 tons; flat-rolling capacity, 2,000,000 tons. Workers employed in October 1965, 8,867.

Siderúrgica Riograndense S.A. Location, Rio Grande do Sul. Founded, 1938. Principal steel producer in Southern Brazil. Has the only continuous casting operation in South America. Privately owned. Has two plants. One, the Rio dos Sinos plant which houses the continuous casting plant and one a small flat-rolling mill, electric furnace, and foundry. Total ingot capacity, 90,000. Nonflat capacity, 75,000. Employs 1,634.

Electro Aço Altona S.A. Location, Blumenau, Santa Catarina. Founded, 1950. Electric steel melt shop. Produces light bars and has facilities to produce steel castings and some forgings. Ingot capacity, 6,000 tons. Privately owned.

Companhia Industrial Itaunense. Location, Itauna, Minas Gerais. Founded, 1911. The firm, family owned, had two departments: a textile and a steel department; in 1963 started to produce steel ingots for sale. Ingot capacity (electric furnace) is 20,000 tons a year. Employs 104.

APPENDIX II

Small Pig Iron Enterprises in Brazil

The post–World War II industrial spurt in Brazil resulted in a substantial increase in demand for pig iron by foundries, semi-integrated steel plants and even integrated firms whose pig iron capacity had not yet reached a level adequate for its steel shop and foundry needs. This situation resulted in the creation of many small blast furnaces established in the state of Minas Gerais. Proprietors of land containing iron ore would build small charcoal-based blast furnaces and sell their products to foundries or medium- to large-scale steel producers. By the mid-sixties these small producers had overexpanded and the pig iron capacity of integrated steel mills was large enough to take care of a large proportion of the industry's pig iron needs. There existed a total of 89 firms with a yearly capacity of about 950,000 tons of pig iron. By 1965, only 30 per cent of this capacity was in use—61 of these small establishments were closed. Below are listed the names of most of these firms.

Firm		Location (by municipio)	Capacity Daily	Capacity Yearly	Employment
Cia. Sid. Pitangui	(1958)[a]	Pitangui	60 t	21,600	123
Cia. Mineira de Sid.	(1942)	Divinópolis	80[b]	28,800	80
Sid. Indust. Mineria	(1959)	Divinópolis	60[b]	21,600	49
Cia. Sid. São Marcos	(1958)	Divinópolis	50[b]	15,000	91
Cia. Melh. Divinop.	(1956)	Divinópolis	25	9,000	44
Cia. Sid. S. João Ltda.	(1959)	Divinópolis	20	7,200	48
Sid. Orion Ltda.	(1959)	Divinópolis	50[b]	18,000	79
Sid. Bandeirante Lt[a].	(1959)	Divinópolis	50[b]	18,000	69
Sid. Divinópolis Lt[a].	(1958)	Divinópolis	20	7,200	49
Sid. St[o] Antonio	(1959)	Divinópolis	20	7,200	39
Sid. São Cristovão	(1959)	Divinópolis	20	7,200	34
Sid. Brasília Ltda.	(1959)	Divinópolis	20	7,200	35
Sid. Progresso Ltda.	(1955)	Divinópolis	20	7,200	45
Sid. Tietê Ltda.	(1959)	Divinópolis	25	9,000	58
Sid. Gafanhoto	(1959)	Divinópolis	20	7,200	45
Ferroeste Industrial	(1959)	Divinópolis	20	7,200	47
J. Rabelo S. A.		Divinópolis	25	9,000	50
Sid. Ipiranga		Divinópolis	50[b]	18,000	75
Sid. D'Olímpia S. A.	(1959)	St. Ant. Mont.	40[b]	14,400	65
Cia. Sid. Lagôa Prata	(1960)	Lagôa Prata	20	7,200	56
Sid. Cajurense	(1960)	Carmo Cajuru	20	7,200	49
Cia. Sid. Claudiense	(1953)	Cláudio	30	10,800	57
Sid. Pinheiros	(1957)	Cláudio	45[b]	16,200	38
Sid. São Paulo	(1961)	Cláudio	25	9,000	58
Sid. São Gonçalo		São Gonçalo	20	7,200	40
Sid. St. Maria	(1956)	São Gonçalo	20	7,200	37

Firm		Location (by municipio)	Capacity Daily	Yearly	Employ- ment
Soc. Sid. Bonsucesso		Bonsucesso	12	4,320	30
Sid. Sudoeste Minas	(1960)	Bonsucesso	30	10,800	59
Sid. Sete Lagôas	(1960)	Sete Lagôas	25	9,000	54
Sid. Noroeste	(1960)	Sete Lagôas	20	7,200	46
Sid. Matozinhos	(1960)	Matozinhos	20	7,200	45
Ind. Bras. da Sid.	(1960)	Matozinhos	60ᵇ	21,600	47
Sid. São Cordisb.	(1961)	Cordisburgo	60	21,600	85
Sid. Mário Pires		Itabira	25	9,000	55
Soc. Ind. Gov. Val.	(1961)	Gov. Valadares	30	10,800	67
Cia. Sid. Gov. Val.	(1963)	Gov. Valadares	60	21,600	84
Us. Sid. N. Srᵃ, Penha	(1961)	Gov. Valadares	40	14,400	62
Cia. Sid. Maravilhas		Maravilhas	20	7,200	33
Sid. Bom Despacho	(1960)	Bom Despacho	20	7,200	53
Sid. Senh. Fátima	(1962)	Bom Despacho	40	14,400	77
Sid. União Bom Despacho	(1961)	Bom Despacho	30	10,800	68
Sid. Brumadinho S. A.	(1959)	Brumadinho	35	12,600	115
Met. St. Antonio S. A.	(1959)	Rio Acima	20	7,200	40
Usina Soledade S. A.	(1961)	Congonhas	25	9,000	34
Parque Met. Augusto Barb.	(1963)	Ouro Preto	12	4,320	30
Cia. Ouro Negro de Sid.	(1959)	Itauna	65	10,800	65
Sid. Itaunense S. A.	(1951)	Itauna	65ᵇ	23,400	94
Fergas Comércio e Ind. S. A.	(1957)	Itauna	15	5,200	47
Sid. Oeste de Minas S. A.		Itauna	50ᵇ	18,000	100
Sid. Itatiaia S. A.	(1951)	Itauna	60	21,600	60
Sid. Frei Leopoldo Ltdᵃ.	(1959)	Itauna	70ᵇ	25,200	102
Sid. Itacolomi Ltdᵃ.	(1960)	Betim	20	7,200	55
Sid. Amaral S. A.	(1960)	Betim	50ᵇ	18,000	65
Sid. Itaminas Ltdᵃ.	(1961)	Itauna	60	21,600	115
Sid. Pedra Negra	(1961)	Itauna	50	18,000	91
Emp. Man. de Aços	(1959)	Itauna	25	9,000	145
Sid. Alterosa S. A.	(1960)	Pará de Minas	40ᵇ	14,400	48
Us. Sid. Paraense S. A.	(1961)	Pará de Minas	45ᵇ	16,200	58
Cia. Brasil. Sider.		Mateus Leme	30	10,800	
Minas Siderúrgica		Betim	50ᵇ	18,000	
Sid. Fernão Dias		Betim	80ᵇ	28,800	70
Sid. Tapajós Ltdᵃ.		Betim	25	9,000	45
Sid. Seb. Itatiaiucu	(1958)	Itatiaiucu	20	7,200	74

Source: From files of the Banco de Desenvolvimento de Minas Gerais, unpublished; also special article in *Correio da Manhã,* Rio de Janeiro, 15 de Agôsto de 1966.

a. Numbers in parenthesis means year of start operations.
b. Means that there are two blast furnaces.

APPENDIX III

Cost of Production Estimates: Two Methods
Part I: Methodology Used by Author in Estimating Brazilian Steel Costs

The estimates below are based on information obtained directly from various types of iron and steel enterprises. Because of the confidential nature of the information supplied, I decided to construct "models" of firms found in Brazil basing the numbers used on averages of several firms. Information on raw material and labor input is based on information for the year 1964. The capital charge is based on estimates found in "The Iron and Steel Industry of Latin America: Plans and Perspectives," in *Proceedings: United Nations Interregional Symposium on the Application of Modern Technical Practices in the Iron and Steel Industry to Developing Countries*, (Prague/Geneva, November 1963, United Nations, New York, 1964). These estimates were calculated at 15 percent of total value of investment: Brazilian steel experts have stated that as a rule ⅔ of total steel investments consist of machinery with a lifetime estimated at 20 years. The other third consists of permanent installations estimated to last 50 years. The depreciation therefore would amount to 4 percent. Given the weight of interest and amortization in Latin American firms, 15 percent would seem resonable.

The capital information has, on occasion, been adjusted according to information available of various feasibility studies of Brazilian steel firms. In the case of capital charges for charcoal blast furnaces, I assumed about half of the capital charge of a coke furnace, since the investment in a coking plant is not necessary and amounts to about half of the value of investment in a coke oven-blast furnace complex. I have included here only my cost calculations for one large integrated firm using an SM (open hearth) steel shop, a steel shop of a large integrated mill using an LD process, and the calculations for the blast furnace of a medium charcoal-using blast furnace. The charges above my direct calculations are based on the recommendations of engineers.

Large Integrated Mill Using Coke and SM Steel-Making Process
Blast Furnace

Raw Materials (Kgs./t):					In US$ (Cr$957 = US$1)
coke	— 656 kg.	× Cr$42.2	= Cr$27,683		
iron ore	— 911	× 4.2	= 3,826		
sinter	— 635	× 8.4	= 5,334		
limestone	— 239	× 4.8	= 1,147		
			37,990		39.70

Manpower (man hour per ton)
   1.00 mh $\times$ Cr\$808 = Cr\$808   .84
Capital Charge   4.75
                                              or
                                         4.35

       Total blast furnace cost per ton of pig iron   45.29
                                           or
                                         44.89

       Assuming this total to represent 90% of cost,
       total all inclusive cost would be equal to   50.32
                                           or

Steel Furnace S.M.   49.88
   Raw Materials:

pig iron	— 746 kg.	$\times$	Cr\$50.0 =	Cr\$37,300	
iron ore	— 109	$\times$	4.6 =	501	
scrap	— 318	$\times$	26.5 =	8,427	
limestone	— 42	$\times$	4.8 =	202	
fluorita	— 0.9	$\times$	33.6 =	302	
				46,732	48.83

   Manpower
     2.0 mh $\times$ Cr\$808 = Cr\$1,616   1.69
   Capital Charge   3.37
                                           53.89

       Assuming this total to represent 80% of cost,
       total all inclusive cost would be equal to   67.36
       if 75%, all inclusive cost   71.85

Flat Rolled Products
   Raw Materials:
   Slabs-1,263 kgs. $\times$ US\$0.07   88.41
   Electric energy (kwh/t) $126 \times$ Cr\$19=Cr\$2,394   2.50
   Capital Charge   17.92
   Manpower
     0.99 mh $\times$ Cr\$808 = Cr\$800   0.83
   plus 10 percent for conversion costs   10.97
                                             120.63

Nonflat Rolled Products
   Raw Materials:
   Blooms-1,067 kgs. $\times$ US\$0.07   74.69
   Electric energy 133 kwh $\times$ Cr\$ 19=Cr\$2,527   2.64
   Capital Charge   14.54
   Manpower-1.00 mh $\times$ Cr\$808 = Cr\$808   0.84
   plus 10 percent for conversion costs   9.27
                                             101.98

Steel Shop of Large Integrated Mill using LD process
Raw Materials:

Liquid Pig Iron-920 kgs × US$0.048	US$44.16
Iron Ore-4 kgs × Cr$4.1 = Cr$16.4	0.02
Scrap-189 kgs × Cr$26.5 = Cr$5,008	5.23
Oxygen-53 (M Nm³/t) × Cr$44.4 = Cr$2,353	2.46
Limestone-5 kgs × Cr$2.3 = Cr$11.5	0.01
	51.88
Manpower	
1.30 mh × Cr$694 = Cr$902	0.94
Capital Charge	2.26
	55.08
Assuming this total to represent 75% of cost, total all inclusive cost would be	73.44
Assuming 80%, all inclusive cost would be	68.85

Note: The information obtained is from firms which are fairly new. If pig iron consumption were brought down considerably and if the cost of pig iron were brought down as blast furnace productivity increases, L.D. steel would be cheaper than S.M. (open hearth steel). E.g., bringing the price of pig iron down to the level of the price of pig iron used in the S.M. shop above (i.e. US$ 0.040 instead of US$0.048) brings the cost down to US$47.72 instead of US$55.08.

Medium Size Integrated Mill: Cost of Pig Iron Production Charcoal-Using Blast Furnace

Raw Materials:

Iron Ore	— 844 kgs	× Cr$ 2.4 =	Cr$2,026	US$	2.12
Sinter	— 626	× 5.0 =	3,630		3.27
Manganese	— 1.8	× 18.2 =	3,276		3.42
Limestone	— 0.5	× 4.8 =	24		0.02
Charcoal	— 3.0	cubic meters × Cr$5,641 =	Cr$16,923		17.68
					26.51
Manpower — 1.39 mh × Cr$776 = Cr$1,086					1.13
Capital Charge					3.40
					31.04
Assuming this total to represent 90 percent of total all inclusive cost, latter would be					34.49

Part II: ECLA COST ESTIMATES FOR VOLTA REDONDA, 1963
  (in US$ per ton)

Blast Furnace (production level: 854,000 tons)

Raw Materials:

Iron Ore	(790 kgs)	US$	2.23
Sinter	(740 kgs)		3.24
Manganese	(25 kgs)		.32
Coke	(855 kgs)		20.24
Limestone	(295 kgs)		1.52
Credit for gas			−3.00
			24.55

Labor	0.73
Other Conversion Costs	3.02
Capital Charge	12.30
	40.60

SM Steel Shop (production level: 970,000)

Raw Materials:

Liquid pig iron	(881 kgs)	US$	33.66
Scrap	(275 kgs)		10.02
Iron Ore	(85 kgs)		.24
Other			3.60
			47.52

Labor		1.70
Other Conversion Costs		14.49
Capital Charge		8.10
	Total cost	71.81

Flat Products (549,000 tons)

Raw Materials:

Steel Ingots	(1,489 kgs)	US$	106.41
Fuel			1.48
Credit for scrap	(360 kgs)		−11.52
			96.37

Labor		2.97
Other Costs of services and materials (of which electric energy is US$3.90)		10.50
Capital Charge		47.40
	Total cost	157.24

Nonflat Products (128,728 tons)

Raw Materials:		
Steel Ingots	(1,265 kgs)	90.26
Fuel		0.95
Credit for scrap	(200 kgs)	−6.40
		84.81
Labor		1.43
Other services and materials		9.00
(of which electric energy 2.24)		
Capital Charge		20.10
	Total cost	115.34

Source: United Nations, CEPAL, *La Economia Siderurgica de America Latina,* Febrero de 1966, pp. 221–227.

APPENDIX IV

Effects on Productivity and Cost of National Coal

The use of domestic coal in Brazil's steel industry has had an adverse effect on both cost and productivity.[1] In 1965 American coal at Usiminas cost US$ 22.05 a ton, while domestic coal cost US$ 42.45. The high price of domestic coal is brought about by the precarious mining conditions in Brazil, the bad quality of the coal (high ash content) which necessitates expensive washing operations, and the inefficient transport conditions to the steel mills.

Besides its high cost, the use of domestic coal in making coke reduces the efficiency of the blast furnace; that is, it increases the coke rate of the blast furnace. Studies have shown that with an ash content of 16 percent, the maximum amount of national coal that could be used in making coke would be 40 percent.

The table below shows the influence on total output and cost of the use of domestic coal in various proportions. It is assumed that all other inputs remain constant. The cost increases show the increase in total cost when one goes from zero to ten percent domestic coal input, from ten percent to twenty percent domestic coal input, etc.

1. A. Lanari, "Consumo de Carvão Nacional na Siderurgia," *Geologia e Metalurgia,* No. 27 (1965).

Effect of Domestic Coal Consumption on the Level of Production and on the Cost of Production of Various Products at Usiminas

% of domestic coal	Pig iron		Steel ingot		Slabs		Heavy plates	
	Output	Increase in cost (US$/ton)	Output	Increase in cost (US$/ton)	Output	Increase in cost (US$/ton)	Output	Increase in cost (US$/ton)
0	—	—	—	—	—	—	—	—
10	446,200	2,736	472,972	2,580	397,296	3,073	305,918	3,986
20	417,400	5,686	442,444	5,362	371,653	6,385	286,173	8,285
30	388,600	8,913	411,916	8,405	346,000	10,009	266,420	12,986
40	359,800	12,680	381,388	11,957	320,366	14,240	246,682	18,475
50	331,000	16,923	350,869	15,958	294,722	19,005	226,936	24,657
60	302,200	21,591	320,332	20,360	269,079	24,247	207,191	31,458
70	273,400	26,701	289,804	25,179	243,435	29,985	187,445	38,903
80	244,600	32,336	259,276	30,493	217,792	36,313	167,700	47,114
90	215,800	38,613	228,748	36,412	192,148	43,362	147,954	56,259
100	187,000	45,726	198,220	43,120	166,505	51,350	128,206	16,623

Notes: Output in tons; ingot steel uses 106 percent pig iron; Lanari's calculations are based on observations made on Usiminas's first blast furnace, which produced 1000 tons a day with 40 percent domestic coal.

BIBLIOGRAPHY

Adams, Walter, and Joel B. Dirlam. "Big Steel, Invention, and Innovation." *The Quarterly Journal of Economics*, May 1966.

American Iron and Steel Institute. *The Making of Steel*, Second Edition. New York, 1964.

Anuário Banas. *Siderúrgia*, 4a. edição. São Paulo: Editora Banas, 1967.

Anuário Estatístico do Brasil.

APEC, Análise e Perspectiva Econômica. *A Economia Brasileira e Suas Perspectivas*. Rio de Janeiro: APEC Editora S. A, six yearly volumes available from 1962 to 1967.

Baer, Werner. *Industrialization and Economic Development in Brazil*. Homewood, Illinois: Richard D Irwin, Inc., 1965.

Banas, Geraldo. *A Indústria Siderúrgica No Brasil*. São Paulo: Editora Banas, 1960.

Banco Nacional do Desenvolvimento Econômico, Departamento Econômico. *Indústria Siderúrgica: Tendencias da Oferta e Procura Globais 1960–9*. Rio de Janeiro, 1960.

Banco Nacional Do Desenvolvimento Econômico, Departamento Econômico. "Mercado Brasileiro de Aço." Rio de Janeiro, 1965, mimeographed.

Bastos, Humberto. *A Conquista Siderúrgica No Brasil*. São Paulo: Livraria Martins Editora, 1959.

Bergsman, Joel, and Arthur Candal. "Industrialization: Past Success and Future Problems," in *Essays on the Economy of Brazil*, edited by Howard S Ellis. Berkeley: University of California Press, 1969.

Boletim dos Custos.

Booz Allen & Hamilton International. "Brazilian Steel Industry Survey." Rio de Janeiro, Banco do Desenvolvimento Econômico, August 18, 1966.

Brisby, M. D. J., P. M. Worthington, and R. J. Anderson. "Economics of Process Selection in the Iron and Steel Industry," *Journal of the Iron and Steel Institute*, September, 1964.

Carvalho, Elysio de. *Brasil: Potência Mundial, Inquérito sôbre a Indústria Siderurgica No Brasil*. Rio de Janeiro: S A Monitor Mercantil, 1919.

Comissão Executiva do Plano Siderúrgico Nacional. *Relatório*. Rio de Janeiro: 1940–41.

Comissão Mista Brasil–Estados Unidos para Desenvolvimento Econômico, *Projetos Diversos;* XIV, Chapter Four, "Companhia Siderúrgica Barbará," 25–94. Rio de Janeiro: 1953.

Companhia Ferro e Aço de Vitória. "Estudo para Escôlha do Local da Etapa de Integracão." Rio de Janeiro, 1962. Study made by J. M. Falcão.

Companhia Siderúrgia Nacional. *Expansão de Volta Redonda, Plano D*. Rio de Janeiro, 1965.

Companhia Siderúrgica Paulista. "Cosipa," "Report for Program for Construction of Steel Works at Piaçaguera for Production of Flat Rolled Products."

Foreign Department, Koppers Company, Pittsburgh, Pennsylvania, February 1958.

CONSULTEC. "O Mercado Brasileiro de Produtos Siderurgicos," Documento 13. Rio de Janeiro, 1961. Mimeographed.

Cotrim, N. C B. *A Indústria Siderúrgica No Brasil e a Necessidade de Formação de Pessoal Técnico para Atender a sua Expansão.* Special pamphlet published by Volta Redonda, 1966.

Dunlop, John T., and Vasilii P. Diatchenko, editors. *Labor Productivity.* New York, McGraw-Hill Book Company, 1964.

Ellis, Howard S. *Essays on the Economy of Brazil.* Berkeley: University of California Press, 1969.

Forbes, Robert H. "The Black Man's Industries," in *Geographical Review,* XXIII, No. 2, April 1933.

Foreign Relations of the United States, Diplomatic Papers, 1940, Volume V, "The American Republics." Washington: United States Government Printing Office, 1961.

Gauld, Charles A. *The Last Titan: Percival Farquhar, American Entrepreneur in Latin America.* Stanford University: Institute of Hispanic American and Luso-Brazilian Studies, 1964.

Gonsalves, Alpheu Diniz, *et al. O Ferro na Economia Nacional.* Rio de Janeiro: Directoria de Estatística da Produção, Secção de Publicidade, Ministerio da Agricultura, 1937.

Hucke, José Bento, and Domingos Esposito Neto. "Defeitos de Aços Nacionais para a Industria Automobilistica," *ABM, Boletim da Associaçao Brasileira de Metais,* Outubro 1960.

IBS, Boletim do Instituto Brasileiro de Siderurgia

International Bank for Reconstruction and Development. "Current Economic Position and Prospects of Brazil," Volume VI, "The Steel Industry." Western Hemisphere Department, Washington, D. C., February 3, 1965. Mimeographed.

Jobim, José. *The Mineral Wealth of Brazil.* Rio de Janeiro; Livraria Jose Olympio, 1942.

Johnson, Harry G. *Economic Policies Toward Less Developed Countries.* Washington, D. C.: The Brookings Institution, 1967.

Johnson, Harry G., *The World Economy at the Crossroads.* Oxford: Clarendon Press, 1965.

Johnson, William A., *The Steel Industry of India.* Cambridge, Mass.: Harvard University Press, 1966.

Kafuri, Jorge F., and Antonio Dias Leite Jr. "Estudo da Localização de uma Indústria Siderúrgica," *Revista Brasileira de Economia,* Setembro de 1957.

Kaiser Engineers International, Inc. "Engineering Report on COSIPA," Steel Mill Project for Cia. Siderúrgica Paulista, at Piaçaguera, Brazil, August 20, 1957.

Kendrick, David A. "Programming Investment in the Steel Industry," Unpublished Ph.D. Dissertation, Massachusetts Institute of Technology, September, 1965.

Labouriau, F. *O Nosso Problema Siderurgico*. Rio de Janeiro: Typ. Besnard Freres, 1924.

Lanari Junior, Amaro. "Consumo de Carvão Nacional na Siderúrgia," *Metalurgia —ABM*, Vol. XXI, No. 93, 1965.

Lanari Junior, Amaro, "O Projeto da Usiminas e sua Justificativa no Planejamento da Siderurgia Brasileira," *Geologia e Metalurgia*, No. 23, 1961.

Leff, Nathaniel H. *The Brazilian Capital Goods Industry: 1929–1964*. Cambridge, Mass.: Harvard University Press, 1968.

Lima, Heitor Ferreira, "Industrias Novas No Brasil—Siderurgia No Passado," *Observador Economico e Financeiro*, ano XXIII, Nos. 264 and 265.

Macedo Soares e Silva, Edmundo de. "Política Metalúrgica do Brasil," *ABM, Boletim da Associação Brasileira de Metais*, Janeiro 1946, Vol. II, No. 2.

Macedo Soares e Silva, Edmundo de. "Expansão da Siderurgia no Brasil," *Geologia e Metalurgia*, No. 20, 1959.

Macedo Soares e Silva, Edmundo de. "Desenvolvimento da Siderurgia no Brasil nos Ultimos Vinte Anos," *Metalurgia—ABM*, No. 86, Vol. 21, 1965.

Maddala, G. S., and Peter T. Knight. "International Diffusion of Technical Change—a Case Study of the Oxygen Steel Making Process," *The Economic Journal*, September 1967.

Máquinas e Metais

Meier, Gerald M. *International Trade and Development*. New York and Evanston: Harper and Row, 1963.

Mello, Geraldo Magella Pires de. "Histórico, Possibilidades, e Problemas da Siderurgia no Brasil," *Observador Econômico e Financeiro*, No. 262.

Metalurgia, ABM

Ministério da Viação, Brasil. *Revisão do Contrato da Itabira Iron*. Rio de Janeiro, 1934.

Ministério do Planejamento e Coordenação Econômica. Plano Decenal De Desenvolvimento Econômico e Social. *Educação, II*, EPEA. Rio de Janeiro, 1966.

Ministério do Planejamento e Coordenação Econômica, Plano Decenal de Desenvolvimento Econômico e Social, EPEA. *Siderurgia, Metais Não Ferrosos, Diagnóstico Preliminar*. Rio de Janeiro, Abril 1966.

Pelaez, Carlos. "The Development of the Basic Industries of Brazil, 1920–1950: A Critique of Public Policy in the Early Stages of Industrialization," unpublished Ph.D. dissertation, Columbia University, 1968.

Presidência da República, Comissão Executiva do Plano do Carvão Nacional. *Usina Siderúrgica de Santa Caterina*. Prepared by ECOTEC, Rio de Janeiro, 1957.

Presidência da República, Conselho do Desenvolvimento. *Relatório Parcial e Preliminar do Grupo de Trabalho Sôbre Siderurgia*, Documento 15. Rio de Janeiro, 1957.

Presidência da República, Conselho Do Desenvolvimento. *Relatório Parcial e Preliminar do Grupo de Trabalho Sôbre Siderurgia*. Rio de Janeiro, 1955.

Presidência da República, Conselho do Desenvolvimento. *Produção Siderúrgica do Estado de São Paulo, Análise das Causas do Decréscimo da Produção de*

Laminados Verificados em 1955, Documento 15, Anexo No. 2. Rio de Janeiro, 1956.

Presidência da República, Conselho de Desenvolvimento. *Programa de Metas*, Tomo III, "Meta de Siderurgia," pp. 15–30. Rio de Janeiro, 1958.

Report Prepared for the Companhia Ferro e Aço de Vitoria, Brazil. "Expansion and Integration of Operation." Prepared by Arthur G. McKee & Company, Engineers and Contractors, Cleveland, Ohio, June 1964.

Revista Latinoamericana de Siderurgia (Boletin Informativo del Instituto Latinoamericano del Fierro y el Acero).

Rogers, Edward J. "The Iron and Steel Industry in Colonial and Imperial Brazil," *The Americas*, October 1962.

Ruist, Erik. "Comparative Productivity in the Steel Industry," in *Labor Productivity*, edited by John T. Dunlop and Vasilii P. Diatchenko. New York: McGraw-Hill Company, 1964.

Sherwood, Frank P. *O Aumento do Preço do Aço da C. S. N.—Estudo de um Caso*. Rio de Janeiro: Fundação Getúlio Vargas, Cadernos de Administração Publica, 61, 1966.

Silva, Raul Ribeiro da. *Indústria Siderúrgica e Exportação de Minério de Ferro*, (third edition). Estudos, Projetos e Proposta, Apresentados ao Governo Federal, Rio de Janeiro, 1939.

Simonsen, Roberto C. *Brazil's Industrial Evolution*. São Paulo: Escola Livre de Sociologia e Política, 1939.

United Nations. *A Study of Industrial Growth*. New York, 1963.

United Nations. *Comparison of Steel-Making Processes*. Economic Commission for Europe, New York, 1962.

United Nations. *Interregional Symposium on the Application of Modern Technical Practices in the Iron and Steel Industry to Developing Countries*, Prague-Geneva 1963. New York: United Nations, 1964.

United Nations, CEPAL. "La Economia Siderurgica de America Latina." Santiago, Chile, Febrero 1966. Mimeographed.

United Nations, Comissão Econômica Para a América Latina. "A Economia Siderúrgica da America Latina: Monografia do Brasil" (prepared by J. M. Falcão). Santiago, Chile, December 1964. Mimeographed.

U. S. Steel. *The Making, Shaping and Treating of Steel*, Eighth Edition. Pittsburgh, 1964.

Veiga, Oswaldo Pinto da. *Producão e Suprimento de Carvão Metalúrgico do Brasil*. Conferência Realizada no Centro Morais Rego Por Ocasião da XV Semana de Estudos dos Problemas Minero-Metalúrgicos do Brasil, Maio 1963.

Veiga, Oswaldo Pinto da. *O Problema do Carvão Caterinese*, Seminário Sócio-Econômico De Santa Catarina, Confederação Nacional da Indústria, Federação das Indústrias de Santa Catarina, Serviço Social da Indústria. Rio de Janeiro, 1961.

Wirth, John D. "Brazilian Economic Nationalism: Trade and Steel under Vargas." Unpublished Ph.D. dissertation, Stanford University, March 1966.

Yearly Reports of Steel firms. Though not listed here, I have made extensive use of the yearly reports of the major Brazilian steel firms.

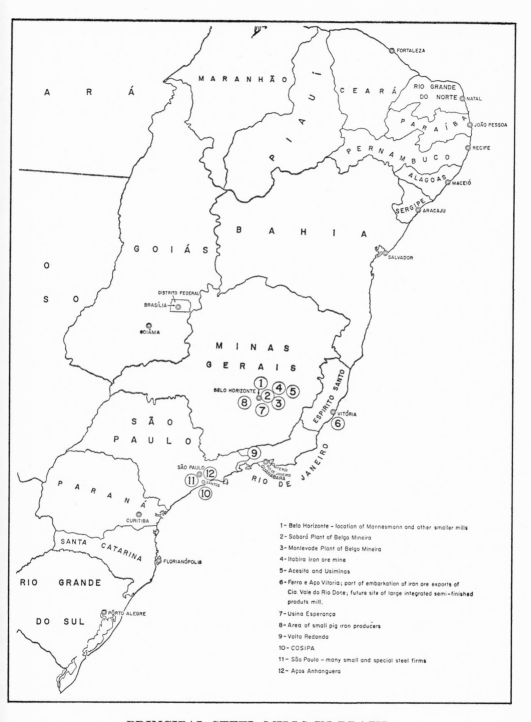

1- Belo Horizonte - location of Mannesmann and other smaller mills

2- Sabará Plant of Belgo Mineira

3- Monlevade Plant of Belgo Mineira

4- Itabira iron ore mine

5- Acesita and Usiminas

6- Ferro e Aço Vitoria; port of embarkation of iron ore exports of Cia. Vale do Rio Doce; future site of large integrated semi-finished produts mill.

7- Usina Esperança

8- Area of small pig iron producers

9- Volta Redonda

10- COSIPA

11- São Paulo - many small and special steel firms

12- Aços Anhanguera

PRINCIPAL STEEL MILLS IN BRAZIL

INDEX